Sir:

IRA BRADLEY & CO., No. 20 Washington Street, Boston, respectfully announce that they have just published

THE DIFFERENTIAL CALCULUS,

With unusual and particular analysis of its elementary principles and copious illustrations of its practical application.

By JOHN SPARE, A. M., M. D. Pp. xix, 244. Price $2.00, mailed free.

[OVER.]

TESTIMONIALS.

A Professor of Engineering in a Scientific School, writes:

"Even a slight examination shows me that the author is tilling a new field. If he succeeds in stripping the Calculus of its terrors, he will do a good service."

A late eminent and venerable instructor for many years, and author of the most known series of Mathematical Works, wrote:

"A work of this kind has been much wanted. It must have cost much time and patience."

A Professor of Mathematics of over thirty years' standing in a New England College, writes:

"I see evidence throughout of original thinking and clear common sense views of the meaning of processes in the Calculus. It is undoubtedly correct, as the author has done, to consider the Calculus, so far as algebraic functions are concerned, as the supplement of common Algebra. I shall read it with interest."

The American Journal of Science and Arts, for March, says:

"The plan of this treatise is different from that of most works upon the same subject. The aim seems to have been to treat the Calculus as peculiarly the continuation of Algebra. Hence, most of the applications are to problems which resemble as much as possible the ordinary problems of Algebra."

The Massachusetts Teacher, of March, says:

"So far as we have read, we have been pleased with the clearness with which a subject generally considered obscure is here presented. We commend it to the attention of our mathematical friends."

The Boston Transcript of Feb. 10th, 1865, says:

"It will be found a valuable aid in Mathematical study; algebraic problems being placed before the learner in the phase in which the Calculus is required for their solution."

The American Literary Gazette and Publishers' Circular, for March, says :

"It gives intelligibility and vitality to the Calculus, without emasculating it of its difficulties. The study indeed becomes attractive under this mode of treatment. The novel feature of practical examples in numerical concrete problems, is peculiar to this work. It is evident that the preparation of the volume has been a steadfast and long continued labor of love with the author, and we think he is entitled to the credit of having made a very important contribution to mathematical study."

The Army and Navy Journal of March 25th, says:

"So far as we know, this is the first instance of a treatise upon the Calculus, in which practical algebraic problems of the sort here profusely presented, have ever been used to illustrate and teach the Calculus. Much interest and entertainment by this course is awakened in a subject from which young students are apt to shrink as dry and difficult."

The Springfield Republican of April 12th, says:

"Mathematicians will be both amused and instructed by it. It contains many novel and curious processes. The work would be obscure if we were not able to catch his idea," etc. . . . . "The mathematicians say that the work is original and valuable."

The author claims that no enthusiasm has been inspired, or success achieved by the use of any previous American books on this subject; and that the "terrors" of the study are inspired only by the ill adapted character of these books to interest and entertain, and by the mass of incorrect statements they contain in whole pages together.

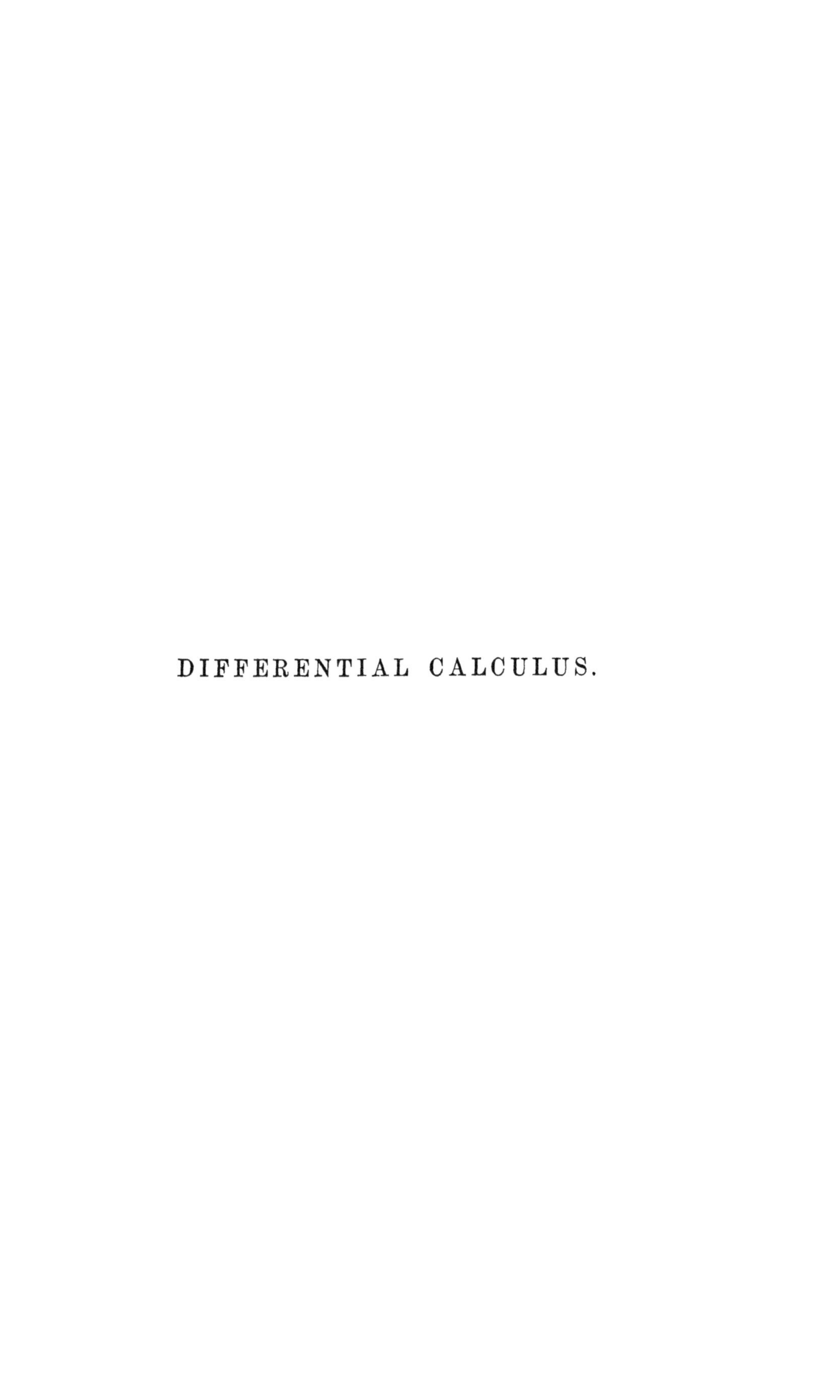

DIFFERENTIAL CALCULUS.

THE

DIFFERENTIAL CALCULUS:

WITH

UNUSUAL AND PARTICULAR ANALYSIS OF ITS ELEMENTARY PRINCIPLES, AND COPIOUS ILLUSTRATIONS OF ITS PRACTICAL APPLICATION.

BY

JOHN SPARE, A. M., M. D.

BOSTON:
BRADLEY, DAYTON AND COMPANY,
20 WASHINGTON STREET.
1865.

In manuscript, by the same author, and the publication contemplated,

THE ELEMENTS OF
DEMONSTRATIVE GENERAL ARITHMETIC.

ELECTROTYPED AT THE
BOSTON STEREOTYPE FOUNDRY,
No. 4 Spring Lane.

PRESS OF GEO. C. RAND AND AVERY.

PREFACE.

The present treatise on the Differential Calculus is believed to be the first, of any character, that has been written and published in America as the special topic of a volume; and the first, so far as known to the author, ever published, that professes the character of the present one.

In our country, and at this day, every mathematical book must be eminently analytical and practical for the many. In European countries, and past times, works upon the Differential Calculus have labored to show, with curtness and severity, just what the science has, at any date, discovered in far-reaching achievement, for the few. Some compilations of the latter have been adopted here, to find which, it is necessary, for the most part, to search works on analytical geometry.

The calculus being algebra, a strictly numerical science, the present treatise claims to have labored successfully in putting on the true character as such. No insinuation is allowed to prevail that it is any part whatever of analytical geometry, or that it is other than the natural sequel and supplement of common algebra; useful, indeed, as an appliance, to borrow, in investigation of the few kinds of geometrical quantity.

Aware of the indispensable importance, to a learner of any new branch of science, that his already acquired knowledge of the most nearly allied character should be adopted as the central principle, around which the new ideas and suggestions are to acquire distinctness and character, the author has commenced this treatise with the terms and appliances of algebra, assiduously preserved and employed. The student is thus enabled to hold his familiar ground, see his former paths and landmarks, find the new objects designed for his attention, tangible and actual, the fruits such that he may grasp them, and add to the previous nurture and furnishings of his mind. Accordingly, he will find here his favorite algebraic problems placed before him in the phase in which the calculus is required for their solution. The author is not aware that concrete, practical problems of this character were ever before published. In this manner is shown the early and elementary nature of the calculus; that it entwines itself around the very threshold of mathematical inquiry.

It is a definite, but perfectly normal fact in the history of science, that the distinguished explorers of the mathematical laws of physical science were obliged to suspend their researches, and come to a stand-still, in order to develop, for necessary use, the elementary principles of the calculus, then unknown to the world. Thus always will a neglect of these, or any elementary truths, by persons who should be well informed concerning them, avenge itself afterwards in their perplexity.

This treatise has been prepared under the strong conviction that its plan of analysis could not fail of adoption at some stage of the natural prosecution of our general modern system of instruction: it is simply the extension to this science, in which an interest is becoming manifest, of just the analytical methods already in use in reference to most other branches of learning, particularly elementary mathematics.

In the many practical problems offered for solution, as a distinguishing feature, the work aims at cultivating and prolonging the enthusiasm of the student, by clothing his conceptions of quantity in the garb of romance, or something of a supposable human experience; these conceptions may, with the more interest, be erratic and fanciful as to economical life, without ever filling or exhausting the generality of pure mathematical conception.

It would not be practicable to present in detail, in this place, the different features of the work. It has been prepared with a great deal of pains, and with reference to a well-considered plan. The consecutiveness of the analysis has been kept in view; not the accommodation of the equal and consistent progress by a learner of a given intelligence, through all its pages, in a given number of weeks. It has rather the character of a hand-book, for collateral use through much of a mathematical course, meeting the different grades of the growing intelligence of a few years, as to one person.

J. S.

New Bedford, Mass., 1865.

CONTENTS.

SECTION I.

ELEMENTARY PRINCIPLES.—A VARIABLE.

SECTION II.

DEFINITIONS RELATING TO FUNCTIONS.—THE USE OF SIGNS.

SECTION III.

THE DEGREES OF AN EQUATION.—FIRST AND SECOND DEGREES.

SECTION IV.

INCREASE OF THE VALUE OF FUNCTIONS.—DECREASE OF THE VALUE OF FUNCTIONS.—STATIONARY VALUES OF THEM.

SECTION V.

THE BINOMIAL SERIES.—SOLUTIONS OF PROBLEMS BY INEQUATIONS.

SECTION VI.

DIFFERENTIAL OF A VARIABLE.—DIFFERENTIAL OF A FUNCTION.

SECTION VII.

FIRST DIFFERENTIAL COEFFICIENT.

SECTION VIII.

USE OF FIRST DIFFERENTIAL COEFFICIENT.

SECTION IX.

SUCCESSIVE DIFFERENTIATION.—SECOND, THIRD, ETC., DIFFERENTIALS.

SECTION X.

TAYLOR'S THEOREM.

SECTION XI.

THEORY OF MAXIMA AND MINIMA.

SECTION XII.

PROBLEMS FURNISHING EXPLICIT FUNCTIONS OF ONE VARIABLE; FOR DETERMINING THEIR MAXIMA AND MINIMA.

SECTION XIII.

COMPLETE HISTORY OF FUNCTIONS.

SECTION XIV.

PRINCIPLES AND PROBLEMS RELATING TO PROJECTED BODIES.

SECTION XV.

THE SYSTEM OF SYMBOLS FOR FUNCTIONS.—AN IMPLICIT FUNCTION OF A SINGLE INDEPENDENT VARIABLE AND ITS DIFFERENTIATION.

SECTION XVI.

PROBLEMS WHICH MAY FURNISH IMPLICIT FUNCTIONS OF ONE VARIABLE, AND CASES OF THEIR MAXIMA AND MINIMA.

SECTION XVII.

FUNCTIONS OF TWO INDEPENDENT VARIABLES: THEIR DIFFERENTIATION AND THEIR MAXIMA AND MINIMA.

SECTION XVIII.

PROBLEMS RELATING TO FUNCTIONS OF TWO INDEPENDENT VARIABLES, AND CASES OF THEIR MAXIMA AND MINIMA.

SECTION XIX.

DEMONSTRATION OF THE GENERAL FORM OF THE DEVELOPMENT OF $f(x+h)$, AND OF THE DIFFERENTIATION OF CERTAIN FUNCTIONS.

SECTION XX.

MACLAURIN'S THEOREM, AND ITS APPLICATION.

SECTION XXI.

DETERMINATION OF THE VALUE OF VANISHING FRACTIONS.

SECTION XXII.

EXCEPTIONAL PRINCIPLE RELATING TO TAYLOR'S THEOREM.

SECTION XXIII.

NATURE OF LOGARITHMS AND EXPONENTIAL QUANTITIES.

SECTION XXIV.

DIFFERENTIATION AND DEVELOPMENT OF LOGARITHMIC AND EXPONENTIAL FUNCTIONS.

SECTION XXV.

EXAMPLES OF THE DIFFERENTIATION AND DEVELOPMENT OF LOGARITHMIC FUNCTIONS.

SECTION XXVI.

EXAMPLES OF THE DIFFERENTIATION AND ANALYSIS OF EXPONENTIAL FUNCTIONS; INCLUDING EXAMPLES FROM COMPOUND INTEREST AND INCREASE OF POPULATION.

SECTION XXVII.

DIFFERENTIATION OF CIRCULAR FUNCTIONS.

SECTION XXVIII.

GEOMETRICAL ILLUSTRATIONS OF THE VALUES OF FUNCTIONS, AND THE CORRESPONDING VALUES OF THEIR VARIABLES; ALSO OF THE VALUE OF DIFFERENTIAL COEFFICIENTS, MAXIMA AND MINIMA, ETC.

DIFFERENTIAL CALCULUS.

SECTION I.

ELEMENTARY PRINCIPLES.—A VARIABLE.

1. THE mode of developing the nature of the Differential Calculus to be adopted in this treatise will be, taking a point of departure within the common principles of algebra; for it is within these principles that the Calculus is based; so far as algebraic quantities are concerned, the calculus is but the completion of omitted algebraic principles — omitted from the general consideration of algebra, for historical reasons only — later and separate invention.

Algebra, in its ordinary character, had omitted to determine a System of Principles according to which the value or resulting amount of a formula must be inferred to change, when a particular component quantity or quantities within that formula should be supposed to increase or to decrease, when near any specific value, or while passing through a range of all possible values. This is the purpose of the Differential Calculus.

2. The following Historical Statement in regard to the Calculus has been derived from Professor Playfair's Dissertation on the Progress of Mathematical and Physical Science:

"Of the new or infinitesimal analysis, we are to consider Sir Isaac Newton as the first inventor, Leibnitz, a

German philosopher, as the second; the latter's discovery, though posterior in time, having been made independently of the former's, and having no less claim to originality. It [the latter's] had the advantage also of being first made known to the world, which was in 1684.

"This infinitesimal analysis, the greatest discovery ever made in the mathematical sciences, as it became known every where enlarged the views, roused the activity and increased the power of geometers, while it directed their warmest sentiments of gratitude and admiration towards the great inventors. By its introduction the domain of the Mathematical Sciences was incredibly enlarged in every direction. Although developed in a state applied to geometry, it was afterwards justly inferred to be independent of it.

"The fluxionary and differential calculus may be considered two modifications [in the matter of notation] of one general method, aptly distinguished by the name of the infinitesimal analysis."

---

3. In preference to the use of abstract language in illustrating the nature of a *variable* quantity, let us use the following problem: —

A fisherman, to encourage his son, promises him 5 cents for every throw of the net by which he shall take any fish, but the son is to remit to the father 3 cents for each unsuccessful throw; after 12 throws they settle, when the father pays the son the amount of the agreement, which proved to be 28 cents; what was the number of the successful throws of the net?

Let $x =$ the number of successful throws.

then $5x - 3(12 - x) = 28$,

that is, virtually, $8x - 36 = 28$,

which is an equation of the First Degree.

4. Here we have, derived from the conditions given,

an expression for the final sum paid the son, which expression would not be different, whatever that sum might have been found in the event to be. The expression

$$5x - 3(12 - x),$$

has a specific value, because it is equated with 28; consequently x is found to have the specific value 8. If there had been a reserve in the problem whereby the amount of the payment had not been declared, or if the formula for payment were to remain for any number of trials of the 12 throws, in accordance with which, 28 cents could hardly be expected to be the sum paid at each settlement, we should still be able to determine all the relations between the *changes* of this sum due the son, and the *changes* of the number of successful throws.

5. If the amount paid or to be paid, were to be styled a sum of money, or y cents, we might look upon the expression

$$5x - 3(12 - x) = y$$

as one in which neither x nor y has a determinate value; but they have exactly determinate *relative* values. If x or the number of successful throws, receive original suppositions of value, y or the formula has an inferred or a relative value:

If $x =$	0,	then $y =$	-36	cents.
" $x =$	$4\frac{1}{2}$,	" $y =$	0	"
" $x =$	14,	" $y =$	76	"
" $x =$	-58,	" $y =$	-500	"
" $x =$	12,	" $y =$	60	"
" $x =$	11,	" $y =$	52	"

It is at once evident that if x be increased by 1, y will be found increased 8 times as much, each of them in their respective kinds of units.

6. There is no *algebraic* limit to this relation of the *change* in y being 8 times the change in x, between infinite positive and negative values of either; nor any interruption in the case of fractional values.

The disability of executing in the problem, fractional and negative values in x, is not one which affects in the least the algebraic expression, or the truth of its most general indications. For when the successful throws are assumed to be 14 (although the whole number is but 12), so that the unsuccessful ones must be algebraically expressed $12 - 14$ or -2, in accordance with which view, 76 cents are paid the son at the settlement, the explanation that the indication is correct, is: for every unsuccessful throw the sum paid *to the son* is algebraically -3 cents; if there be -2 of such throws, then

$$-3 \times -2 = 6;$$

now 6 is the number of cents by which

$$5 \times 14, \text{ as found}$$
$$\text{in } 5 \times 14 - 3\,(12 - 14) = 76$$

is properly increased. So that the unlimited amount which may be indicated as payable to the son at a settlement, logically agrees with a correspondingly unlimited number of successful throws, which may be supposed, as indeed it ought.

7. In the really strict use of common language, the limits for the sum paid to the son at a settlement, are from 0 to 60 cents. In the language of algebraic equivalents, there are no limits whatever, in connection with the supposed problem, because its indeterminate quantities so far enter into an equation of the First Degree.

The first member of the equation

$$5\,x - 3\,(12 - x) = y$$

expressing the amount payable to the son in form and

detail, constructed with the indeterminate quantity x, in connection with other determinate quantities, so that the expression varies when that quantity x may, is an instance of a *function of a variable*, which variable x may be. Using y as the *equivalent of the function in amount only*, we may call y that function, when thus equated.

8. A *variable* is a quantity which may have different values, and is represented by the late letters of the alphabet, x, by y or by z.

9. A *constant* quantity, whether known or unknown, is one which by original assumption is not to vary during an investigation into which it enters, or by inference is found to be so conditioned as not to vary, and when known, it is represented by number, as 1, or 20, or by a or b, etc.; d, however, is used to signify *differential*, and its use purposely avoided as any quantity of itself.

10. When several quantities are so related as to determine one another, or to depend on one another by equation; a variety of mutual investigations may be instituted among them, by arbitrarily assuming an *independent* variable or variables, and then examining the law of the variation of the *dependent* variable.

11. Recurring to the problem of the fisherman, we have the equation

$$5x - 3(12 - x) = 28,$$

which answers its algebraic purpose of determining one specific value for x, which is 8, the number of successful throws of the net. If 8 be supplied in the place of x, and any one of the other quantities of the equation were left unstated in the conditions, and *be made* x in a *new investigation*, it could be determined. Indeed if any two of the quantities were unstated in amount, the mutual dependency of the two becomes evident. For a conventional reason we will call that quantity x, to which we may wish to reserve the right to assign arbitrary values, to the extent that we can do so, and will call the other, which receives inferred values, y. The exchange of x for y we will call the *converse* of the investigation. When y can be isolated as one member of the equation, not occurring in the other, the function of x is called *explicit.* We will suppose the *explicit* form of the functions of one variable x, *implied* in some of the following statements of the above equation, to be worked out, and we will suppose their significance to be enunciated in words.

1. $5x - 3\,(12 - x) = y.$
2. $x \,.\, 8 - (12 - 8) = y.$
3. $5 \times 8 - 3\,(x - 8) = y.$
4. $5 \times 8 - x\,(12 - 8) = y.$
5. $8y - x\,(12 - 8) = 28.$
6. $8y - 3\,(x - 8) = 28.$
7. $5x - y\,(12 - x) = 28.$
8. $5 \times 8 - y\,(x - 8) = 28.$
9. $5y - 3\,(x - y) = 28.$
10. $x \,.\, y - 3\,(12 - y) = 28.$

Each of the above has evidently its converse.

It may be useful to adopt a comprehensive enunciation of the problem.

In the parentheses ( ) which follow in the problem, let the following reading be understood: the variables now being supposed to be two, one depending on the other, viz:

(For such number is exactly compatible with the other numerical quantities, received as given, without regard to the brackets.)

In the brackets [ ] let the following reading be understood:

[Or an indefinite numerical quantity, if in one of the other brackets, another indefinite numerical quantity be understood to be read.]

12. A fisherman promises his son 5 cents ( ) [ ] for every throw of the net by which he shall take any fish, but the son is to remit to the father 3 cents ( ) [ ] for every unsuccessful throw; after 12 throws ( ) [ ] they settle, when the father pays the son the amount of the agreement, which was 28 cents ( ) [ ], there proving to have been 8 ( ) [ ] successful throws. Required all the truths dependent on the various readings, for each two of these indeterminate but mutually dependent quantities.

All of the above equations and their converses, except the 8th and 10th are of the First Degree, and the variables in them may have any algebraic values, positive, negative, or infinite, and they vary at uniform rates, in passing from one value to a succeeding one.

In the 8th and 10th and their converses, since the variables are factors together, the equations are of the Second Degree, and their variations are subject to other laws.

13. Such relative rates of variation are subject to exact numerical determination, which it is the object of the succeeding Sections to explain in their general nature.

14. It is now evident that those problems in algebra which are based upon Simple Equations or those of the First Degree, are such only with reference to the one investigation for which they are offered, and that a change of the investigation for determining other relations between quantities fully conditioned in such problem, is likely to require the use of Affected Equations, which are always as high as the Second Degree, and may be higher.

SECTION II.

DEFINITIONS RELATING TO FUNCTIONS. — THE USE OF SIGNS.

15. An *explicit* algebraic function of a variable quantity, is an indicated mode in which addition, subtraction, multiplication, division, and other arithmetical processes with quantities, either, any or all of them, in or among which, said variable is somehow concerned, are used for working out a resultant quantity, and among which indicated operations, this variable quantity holds a marked position, as the one to which a particular reference is to be made in regard to its changes of value.

16. Hence a function is primarily and always a *mode of constituting a quantity;* although a function of a variable may, after some hypothesis for itself or that variable, *possess a value*, the function cannot be considered as merged in a specific amount of quantity. We cannot speak of a *great* or of a *small* function, as some writers do because that means a *great mode*, — a *mode* is not a value or a quantity.

17. Strictly speaking, a function of a variable, having as it may, constant quantities in it, is a function of all the distinct quantities in it, because it requires them all to *perform the office* of representing some quantity in mode, but custom has sanctioned calling the whole expression, that determines some designed quantity, in which the variable occurs, a function of the variable, although the variable occurs in no more than in one of the terms of such expression, and in the simplest manner; and however complicated other terms containing constants may be.

18. Functions which are not *algebraic* have restrictions to Geometry, Trigonometry, or to Logarithms, which are to be considered in this treatise. Functions which are not *explicit*, may be *implicit*, or involved in an equation, and determinate as a *mode*, only after algebraic process. The implicit is the most general and comprehensive. It is often subject to difficulty of solution, or to impossibility.

19. A *formula* is an expressed mode of operating with quantities, for deducing the amount of another. In a formula, as such, there is not necessarily a quantity, to which any reference is to be made as subject to change of value.

An explicit function is a formula, in which there is a quantity, subject to a change of value, and subjecting the formula to a change of value.

20. There is no absolute need of any equation in the statement of an explicit function. The equating of it with its own correspondingly variable amount called by another name y, is often only a piece of convenience. Whenever a function of a variable appears equated with a determinate or constant quantity, supposition is evidently applied for the value, or values if there should be more than one, of that function. If that constant is not removable, or open to supposition for a change of value, the variable x may take the name of unknown quantity; although, since it is determinate it might be regarded as known, in the calculus. We sometimes use a function of a constant; this is when such constant occupies only temporarily the place of a well understood variable.

21. When the conditions of an algebraic problem are stated in the form of an equation, the members of such equation may each be functions of the unknown quantity, but may be reduced to one function, for the value of some other quantity.

22. A marketman purchased fowls: some 2 for a dollar, and as many more 3 for a dollar, and sold them at the rate of 5 for 2 dollars, losing 4 dollars by the operations; required the number of each sort.

Let $x =$ the number of each sort;

then we have these two expressions for one sum of money, which may be put equal, viz:

$$\frac{x}{2}+\frac{x}{3}=\frac{4x}{5}+4.$$

In case the number of each sort or x be variable, these two expressions will permit the forming of one function of x, *for the sum lost*, viz:

$$\frac{x}{2}+\frac{x}{3}-\frac{4x}{5}=4.$$

This function remains the same if the sum lost were indefinite, and were to change only on change of the number of each sort; in the supposition it is, when without an equation:

$$\frac{x}{2}+\frac{x}{3}-\frac{4x}{5}$$

and it is ready for the comparison of *the number* of dollars lost, during changes of the *number of each sort.* It will be convenient to equate it with y, as being also the number of dollars lost.

23. The problem can be enunciated, in the general aspect, for algebraic determinations, and for the calculus determinations, as before, with the fisherman problem.

---

24. All strictly algebraic quantities in the Differential Calculus, are subject to algebraic expression, and are numerical in their nature, and are real or imaginary or irrational in value. All functions have *numerical* amounts, or expressions, for their values,—when rendered determinate and real.

25. Three triangularly placed points ($\therefore$) are used to signify *hence* or *therefore.*

26. One of these characters, $>$ or $<$, is placed between two unequal quantities, the larger quantity of the two being embraced by the limbs of the character, and may be read, — being greater than, — being less than, — or, is greater than or less than.

27. Zero, or naught, is freely treated as a determinate value for quantity. Hence, in the expression $0 > a$, a is negative. A quantity that is negative is freely called less than nothing, for the uniformity and brevity of the mode of expression. When zero forms one member of an equation, it both marks and simplifies it, for some general uses ; hence its use.

28. A single point (.) is used as an abridgment of the sign of multiplication $\times$. When the point is placed directly between figures, it ought to indicate decimals, with nevertheless, easily understood exceptions.

29. A prostrate figure of ∞ signifies an infinite quantity.

30. The sign of equality, $=$ may often be advantageously read as a verb *equals*, sometimes as *equalling*, or as, *that is*, or as, *that is to say*.

31. The Binomial Theorem should be mentioned as the Foundation of the essential principles of the Calculus, and is demonstrated in most treatises of algebra, as related to indexes being whole and positive numbers.

The extension of its demonstration, to embracing binomials having fractional, negative, or imaginary indexes is commonly made in the Calculus as a sequence to Maclaurin's Theorem.

The following ocular views of one application of the Binomial Theorem, will impress its law more concisely than the use of n, $n-1$, $n-2$, etc., will do.

$$(a \pm b)^4 = a^4 \pm \frac{4}{1} a^3 b + \frac{4.3}{1.2} a^2 b^2 \pm \frac{4.3.2}{1.2.3} a b^3 + \frac{4.3.2.1}{1.2.3.4} b^4.$$

32. The differences in the successive supposed values of a quantity, may obviously be fractionally minute; the conceiving of a quantity at such successive values, gives rise to the idea and expression, of the *growth*, *decrease*, etc., of the quantity. It was this unnecessary transfer of the mode of apprehending quantity in this state by the mind, to the quantity itself, that gave rise to considering such quantities as "generated by motion," — "the quantity thus generated is called the *fluent* or *flowing* quantity," — "the velocities with which flowing quantities increase or decrease at any point of time, are called the *fluxions* of those quantities at that instant." (Vince's Fluxions.) These forms of expression have been wisely discontinued; although a modified form of these expressions is sometimes convenient, such as *growth* —*faster* — and *slower*, of the value of a function, or of a quantity.

SECTION III.

THE DEGREES OF AN EQUATION.—FIRST AND SECOND DEGREES.

33. An algebraic equation of the First Degree containing *one* quantity called unknown (but which, however, may be entirely determinate, having a fixed value) in the form best adapted to exhibit its degree, and consequently to offer the best opportunity for showing the nature of its quantities, is by general expression

$$Ax + B = 0,$$

in which A embraces the algebraic aggregate of all given quantities, which are factor with x, and B represents generally all other quantities, and each term has the sign $+$

(plus) in the sense that is understood to embrace minus, in case the particular conditions require it. It is evident that the value of x is determinate when A and B are not each at once zero or infinite.

PROBLEMS.

34. It is required to determine whether the following equations are of the First Degree, and in what form A and B are represented.

1. $$c\,\frac{(a+\sqrt{b})}{x}+c=150\,a.$$

2. $$\sqrt{x}-c\,b=\sqrt{(x-b)}.$$

35. The general equation of the First Degree containing *two* quantities, x and y, indeterminate in value is,

$$A\,x+B\,y+C=0,$$

in which A is the aggregate of given quantities, factor to x, and B of given terms, factor to y, and in which C is the aggregate of terms given separately. Only the quantities implied by A, B, and C being known, x and y cannot be determined by them. Nevertheless it is evident that there is a relation always implied between the values of x and y, which the equation and the fixed values of A, B, and C preserve.

36. In the particular case in which A or B should have the value zero, x or y receives a fixed value. It is intended that A, B, and C should represent determinate and unchangeable quantities in any particular use, and should be general only for the expression of a general formula. In the particular cases in which A should be infinite, while B and C are not, x loses general values, and must become zero. In the particular case in which B should be infinite, while A and C are not, y loses general values, and must become zero. In the particular case when C is infinite, while

2

A and B are not, both x and y may be infinite, and one of them must be infinite.

37. In the particular case when A and C are each infinite, while B is, or is not infinite, since x becomes

$$x = -\frac{By}{A} - \frac{C}{A},$$

it will be indeterminate.

The quantities of which A, B, and C are the representatives, if definite in value may be called *constant* with reference to any investigation, for which a purpose is subserved in making them so, while x and y may *vary* in value, but evidently with a dependence upon each other; they may be *variables.*

38. THEOREM. *In an equation of the First Degree between two variables, if one variable be supposed to change uniformly in value from any supposed definite value, the other must change uniformly in value.*

39. The variables in an equation of the First Degree may have the utmost range of values.

40. But in reference to the *amount* of variation, that variable x, which is factor to A, will vary uniformly as many times y (the factor of B), as is expressed by the quotient $\frac{B}{A}$, and they need not concur in increase or decrease, as the *mode* of simultaneous variation.

41. In the particular case when $A = 1$ and $B = 1$, and y is negative, we have the principle of Simple Addition in Arithmetic, or

$$x + C = y,$$

when it is evident that if x be increased, y is dependently increased. Because, if one of two numbers which are proposed to be added, be first increased, the amount is as much increased. If x be negative, the equation of the First

Degree as above, becomes the representative of the nature of Simple Subtraction in Arithmetic, with a properly corresponding result.

42. In the particular case when A or B is negative and $C = 0$, we have

$$\frac{Ax}{B} = y,$$

$\frac{Ax}{B}$ being like two factors in Simple Multiplication, and y their product, and it is evident that if x change in value, y or the product will change $\frac{A}{B}$ times as much as the one factor x. But it is evident that $\frac{A}{B}$ may be less than a unit.

43. The general equation of the Second Degree between two variables x and y is

$$Ay^2 + Bxy + Cx^2 + Dy + Ex + F = 0,$$

where in the particular case of $A = 0$, $C = 0$, $B = -1$, $D = 0$, and $E = 0$, we have

$$xy = F,$$

which is the representative of the nature of Simple Division in Arithmetic, if we propose to let a divisor and the quotient vary, and the dividend remain unchanged.

44. Here we may expect some law of mutual variation between x and y, different from what appears in respect to them, in equations of other degrees.

If we take any number, as 257, as dividend, and divide it by any number, as 18, and then let the divisor be changed by a specific amount, 3, we have

$$257 \div 18 = 14\tfrac{5}{18},$$

$$257 \div 21 = 12\tfrac{5}{21};$$

and again, divide it by any number, as 8, and by $8+3$, we have

$$257 \div 8 = 32\tfrac{1}{8},$$

$$257 \div 11 = 23\tfrac{4}{11},$$

where we soon perceive that the change in the quotient is not uniform; and we should find that according as the divisor and quotient approach equality with each other, will the rate with which the larger decreases, diminish with a retarding rate, compared with the less, supposed to vary with an assumed uniformity, which is both a possible and necessary mode of making the supposition.

45. In the particular case where $A=0$, $B=0$, $C=1$, $D=-1$, and $F=0$, we have

$$x^2 = y,$$

i. e., $$x = \sqrt{y},$$

in the examination of which we shall find that the second or square roots of a quantity, do not vary uniformly with the quantity.

46. The complete discussion of the equation of the Second Degree, is comprehensive of a protracted variety of particular cases and principles.

In the use that is to be made of *functions of variables*, we may not be able to know all of them by classification as of any numerical degree that has been numbered or defined, although quite desirable when it may be readily determined.

47. When the variables x and y in an equation are at powers expressed by whole and positive indexes, the numerical name of its Degree is determined by the highest sum of the indexes of the variable or variables which are direct factors in any one term, with A or B, etc.

PROBLEMS.

48. It is required to determine of what Degree are the equations:

1. $$\frac{cx}{y} + \frac{y}{ax} - e = b + a.$$

2. $$\sqrt{abx + y^2} = y^2 + a.$$

3. $$\frac{1}{ax - 2} = \frac{b}{(a + b)xy}.$$

The general equation of the Second Degree

$$(1.) \quad Ay^2 + Bxy + Cx^2 + Dy + Ex + F = 0,$$

when y is equated with its value, gives us

$$(2.) \quad y = -\frac{Bx + D}{2A} \pm \frac{1}{2A}\sqrt{\left\{(B^2 - 4AC)x^2 + 2(BD - 2AE)x + D^2 - 4AF\right\}}.$$

Now the quantities represented by the capital letters, are general in these equations, but become particular, in a given applied case. According to the relative value of these constants, a classification of the nature of the Systems of values of y and of x may be made. Of these, three systems of values for y and x are prominent, conditioned thus: they are,

1. when $B^2 - 4AC < 0$, Elliptical;
2. " $B^2 - 4AC > 0$, Hyperbolic;
3. " $B^2 - 4AC = 0$, Parabolic.

49. Since the general equation is symmetrical with reference to y and x, these variables have *corresponding* systems of values, which values are equal when the constants

$A = C$ and $D = E$, and retain a similar nature when these constants are not equal.

50. In the *elliptical* condition, both x and y have a maximum and a minimum value.

51. In the *hyperbolic*, there are two systems of symmetrical values for each variable, isolated from each other; and if there be a maximum for one of them there must be a minimum also for it, and there may or may not at once be a maximum and minimum also for the other.

52. In the *parabolic*, there must be a maximum or minimum for one of the variables, and there may be for the other.

53. In the general equation there may be only imaginary values to one of the variables, when there must also be to the other; these occur when the quantity under the radical sign is found negative in equation (2.)

In geometry much importance is attached to the indications of these equations, in the study of Curves of the Second Degree. The terms, *maximum* and *minimum*, will be seasonably defined.

SECTION IV.

INCREASE OF FUNCTIONS. — DECREASE OF FUNCTIONS. — STATIONARY VALUES OF THEM.

Exercises.

54. 1. There are purchased 99 pounds of a commodity at a price fixed to-day, but liable at the next purchase to be found changed. How many times more will the value of the 99 pounds vary, than the value of a single pound?

2. How many times more will the value of 17,000 pounds change than the value of 1 pound of it?

3. How will the value of one sixteenth of a pound vary compared with the value of one pound of it?

4. How will the capacity of 100 equal casks vary compared with the capacity of one, if the casks should be made larger, on a repetition of the manufacture?

5. How will $b x$ change if x change?

6. How will $(a + b - c) x$ vary in amount if x grow greater?

7. If h expresses the amount that x, in the above two expressions containing x, changes, what expresses the amount of the expression's change?

8. When a commodity is worth 10 cents a pound, and it is quoted as tending upward in price, does the value of 100 pounds *commence* to increase any faster or slower than it would if the price were 15 cents per pound, and the value of the 100 pounds commenced to increase from that? That is, does the *ratio* of the growth of $100 x$ to that of $1 x$ depend on the greatness of the quantity x, or on the amount of the growth of x?

9. Does that function of x, which $100 x$ is, grow uniformly, on uniform increase of x?

10. Will $8 x^2$ grow uniformly on growth of x?

11. How great could the value of 1542 pounds of a commodity ever become, by an indefinite increase of the price of one pound?

12. If x varies, which of these two functions of x, viz., $10 x$ and $11 x$, will vary the more in amount? How much more?

13. If the value of 71 pounds of a commodity be negative, or be considered subtractive from some greater sum, would an *increase* of the value of the 71 pounds be negative?

14. How would the amount of the *decrease* of the value

of the 71 pounds stand, as affecting the original greater quantity?

55. Both the number of pounds of a commodity and the number of cents per pound, will now be conditioned to increase; if there are u pounds, at x cents per pound, and the pounds increase by any amount k, and the cents per pound increase by any amount h, what expresses the new entire value? What expresses simply the increase of value?

1. If the amounts k and h be considered alike, or each h, how does the expression of the increase of the value of the commodity become united?

Ans. From $ux+uh+xk+hk$ to $ux+(u+x)h+h^2$.

2. How does the expression still more condense, if the original number of pounds, and number of cents per pound, were alike, or each x? Ans. Into $x^2+2xh+h^2$.

3. What are the factors of just the increase?

Ans. h and $2x+h$.

56. Thus far the quantities u and k x and h may be any whatever. But a quantity situated and treated like h as an increment of x, *may be* moulded to a purpose, — a merely mathematical purpose, — be a creature of supposition, subservient to the illustration of a mere addition to x, and may be indefinitely small; while x is a quantity to be accepted as given; it is any quantity, till made determinate by some condition.

57. If there are x pounds of a commodity at x cents per pound, and x be increased an indefinitely small amount h; by how many times as much as h, *at the least*, will the value of the whole of this commodity, *commence* to increase?

Ans. $2x$.

How many, to decrease, if x diminish as much?

Ans. $2x$.

1. If the number of pounds be restored as u, while k

may be h according to the above indication, how many times as much as h, *at least*, will the value of the whole commodity *commence* to increase? Ans. $x + u$.

2. How many, if u increase while x diminishes?

Ans. $- x + u$.

3. How many, if u decrease and x increase?

Ans. $x - u$.

4. If x pounds of a commodity, worth x cents per pound, are possessed by each of 9 persons, how will the whole value *commence* to decrease, if x decrease by (an amount that is merely nominal) h? Ans. $9 \times 2x$ times h.

5. If one of these persons obtains his share from a lot of 20 pounds of the commodity, all of the above value, how many pounds does he take when he leaves the greatest possible value possessed by that remaining lot?

6. In $20x - x^2$ apply successively 1, 2, 3, etc., for x, till the greatest remainder be found.

7. If a piece of ground conditioned to be kept square, is to be enlarged, on how few of its sides must the new area *commence* to form? On how few, if the square must decrease, must the deductive area *commence* to form?

8. There is a rectangular piece of ground 10 rods in one dimension; its other dimension is the same and to be kept the same, as the side of a certain square piece of ground; upon the suggestion that the sides, one of each lot which are alike, commence to increase, which lot commences to increase the faster in its area when that common side is 4 rods? When 6 rods? When 2 rods? When 1000 rods?

9. What number of rods must that common side contain, when the greater rapidity of the growths of the two lots, passes from one to the other? From which, to which?

10. When the side of the square lot is 50 rods, and the lot commences to increase, will the product expressing the increase of area, have one of its factors greater than

if the side was 49 rods, and the increase then took place?

11. If we have this function of x, viz., $92x - x^2$, consisting of a minuendive and a subtrahendive term, must the value of the function, which evidently consists of their difference, always increase while the minuendive term increases the faster of the two?

12. Must its value decrease when the subtrahendive term increases the faster? Can a value of x and of the function be found, when the increase of the subtrahendive term, being negative, is equal to the increase of the minuendive term, which is positive, so the sum of the increases equals 0?

Problems.

58. 1. Let it be required to divide the number 92 into two such parts, as when multiplied together, shall produce the greatest product.

$$\text{Let } x = \text{one of the parts.}$$
$$\therefore 92 - x = \text{the other.}$$
$$\therefore 92x - x^2 = \text{their product.}$$

When the function is the greatest in value, it is evident that either adding or subtracting any small and indefinite amount h, to or from x, diminishes the value, so that

$$92(x + h) - (x^2 + 2xh + h^2) < 92x - x^2,$$

and $$92(x - h) - (x^2 - 2xh + h^2) < 92x - x^2;$$

therefore in accordance with what has been said,

$$92h < 2xh + h^2$$

and $$92h > 2xh - h^2,$$

where all the signs of the last expression have been

changed, because the members of the inequation are compared only with each other. Hence,

$$92 < 2x + h,$$

and $$92 > 2x - h;$$

hence, $x = 46$ within one half of the smallest quantity that can be conceived of.

2. A boy having 16 equal parcels of marbles, gave to as many boys as there were marbles in a parcel, as many marbles each as there were boys, and retained the greatest possible number himself, for any number in a parcel. Required the whole number of marbles; of parcels; of boys; number of marbles given away, and the number retained. Ans. 128, 8, 8, 64, and 64.

3. Some benevolent persons, gave each 240 dollars to some orphans; four times as many orphans each month received 2 dollars each, and during three times as many months as there were contributors; the unexpended sum was still the greatest possible. How many persons gave? Ans. 5.

59. It is *generally* possible, when the terms of a function are simple and *few*, not fractional, when the indices are positive and entire, to foretell, on logical principles, whether it is a maximum or minimum that a function has, if but one. For instance, by noting the sign of the term having the variable at the highest positive power, we shall perceive that at some positive value of the variable, such term will rule; if this be positive the presumptions are, a minimum at some value of the variable, if there be either, — if this be negative the presumption is reversed; such term being destined to infinity in itself, will leave the maximum or minimum behind.

4. Consider $x^2 - 92x$ in these regards.

Infinite and positive are easily inferred; a minimum probable. If the function be considered settled at a minimum, then x takes a definite value; and if x be increased, the positive member must increase more than the other diminishes; so that, as before,

$$2xh + h^2 > 92h;$$

and again retrospectively, subtracting the last that x gained in its growth, for the function by hypothesis did diminish, and the negative term had diminished more than the positive increased, so that we have, as before,

$$92h > 2xh - h^2,$$

and $$x = 46 \text{ as before.}$$

So it appears that nothing is different in discovering a minimum, from discovering a maximum.

The consideration of a function, when not readily disclosing its tendencies in these respects, especially when it probably has both maxima and minima, is deferred to another section of the treatise.

5. From a cistern holding 6218 times a certain measure full, were taken away that measure full 11 times each day, during as many days as that measure held gallons. The greatest possible number of gallons were left in the cistern at last; then how many quarts did that measure hold?

Ans. $1130\frac{6}{11}$.

6. A grain dealer sells to A the same number of bushels of grain out of 100 bushels, as he sells the remainder to B for, in cents additional to 25 cents per bushel, and realizes the greatest possible sum from that sold to B. Required the number of bushels sold to A, and the price per bushel of that sold to B.

Ans. $37\frac{1}{2}$ bushels, and $62\frac{1}{2}$ cents.

7. From a cask containing 19 times a certain measure

full of water, a smaller measure by 7 quarts is drawn full, as many times as it held quarts, when the greatest possible quantity of water remained behind in the cask. Required the capacities of the measures. Ans. $16\frac{1}{2}$ and $9\frac{1}{2}$ qts.

8. From a lot of 1501 (a) bushels of hay-seed 79 (b) casks full were put up for exportation; the remainder were sold for home consumption, at 3 (c) dollars per bushel more than there were bushels in a cask. Required to adjust all the quantities which are not stated definitely, to the greatest value of the home sale.

Let $x =$ number of bushels per cask.
$\therefore x + c =$ dollars per bushel.
$\therefore bx =$ number bushels exported.
$\therefore a - bx =$ number bush. sold at home.
$\therefore (a - bx)(x + c) =$ number dollars home sale.
or $(a - cb)x - bx^2 + ac =$ number dollars home sale.

When this last sum is greatest if x be increased, and afterwards diminished by h, we have,

$$(a - cb)h < 2bxh + h^2,$$

$$(a - cb)h > 2bxh - h^2,$$

and $$\therefore x = \frac{a - cb}{2b} = \frac{1501 - 3 \times 79}{2 \times 79} = 8.$$

Amount of sale \$9559.

It will be perceived that the quantity ac, being constant, disappears from the reasoning. In the generalization, if $cb > a$, the value of x would be negative.

9. A boy was offered the use of a rectangular playground, which he could surround with his kite string, 80 rods long, but he must determine its dimensions such, should he afterwards incline to increase or diminish the width the least, it would be *at the rate* of enlarging or diminishing the area, 10 square rods for one rod that

should be added to the width of the playground, or taken from it.

Let $x = \text{width}.$

$$\therefore 40 - x = \text{length}.$$

$$\therefore 40\,x - x^2 = \text{area}.$$

Now, if x become $x + h$, $40\,h - 2\,x\,h - h^2$ is the area's increase; but it is conditioned to be equal to the area of h by 10 rods, i. e., when $h = 0$, but when $h > 0$ we must have,

$$10 > 40 - 2\,x - h.$$

Likewise, if x become $x - h$, the area's diminution in amount when positively expressed, will be $40\,h - 2xh + h^2$, which also $> 10\,h$ by the same condition; hence,

$$x > 15 - \tfrac{1}{2}\,h,$$

and $$x < 15 + \tfrac{1}{2}\,h;$$

which conditions make an equation, when h begins to be an amount, and while it is practically 0; $\therefore x = 15$. In some sense h may be considered a mere suggestion of quantity.

10. How great could the playground ever be made? How small? Does it change faster when near a square, or when most at variance with a square?

11. The number of pounds of a certain commodity added to its number of dollars value per pound is 576; what is its price, when the suggestion that the price begins to vary the least (in our mental adjustments of it), requires the inference that the value of the whole will begin to vary 19 times as fast?

12. A very cautious man has been offered the opportunity of laying out for himself a rectangular piece of ground, which shall contain 6 acres; two contiguous sides of which must agree with, and be a part of an established north and south, and an east and west line, meeting like two edges of the sheet of this page. The particular difficulty has been the

establishing of the unspecified corner. He has walked over the ground line so much in which he might establish said corner, that he has worn a path a portion of the way. He could not walk over the whole track, because it may be found to be infinite in length. He finally determines the boundaries so that the lot shall embrace a certain spring of good water, and drives down his stake for the indeterminate corner at such a point in that path, that the suggestion of varying it along that path, the least amount, requires the inference that the length of his lot must increase $3\frac{1}{2}$ times as fast as the width would diminish. Required the length and width.

13. In the equation $xy - y - a = 0$, a being a constant x and y variables; if x can be 22, what will be y, and what the tendency and rate of y to increase or decrease, on increase of x (if it can increase)?

Ans. y to decrease $2\frac{252}{289}$ times as fast as x to increase.

SECTION V.

THE BINOMIAL SERIES. — SOLUTIONS BY INEQUATIONS.

60. THEOREM. *If the quantity in the Binomial Series, whose powers increase, in the successive terms, be sufficiently small, any term of such series will be greater than the sum of all that follow it, or less.*

If the binomial $(x \pm h)^n$ be developed by this series, n being a whole number, negative or fractional, it becomes,

$$x^n \pm Ax^{n-1}h + Bx^{n-2}h^2 \pm Cx^{n-3}h^3 + \text{etc.};$$

that is,

$$x^n \pm (Ax^{n-1} \pm Bx^{n-2}h + Cx^{n-3}h^2 \pm \text{etc.})\,h:$$

where A, B, and C are quantities which contain neither x nor h, but are coefficients which may be otherwise composed of n and numerals.

61. Now, it is evident that the first term within the parenthesis will not grow smaller by any diminution which h may undergo, as all of the succeeding terms do without limit, and consequently their sum, and become less than any assignable quantity, and, of course, less than that first term.

The above is evidently true whether n be negative or fractional, because we have only to regard x, or its reciprocal $\frac{1}{x}$, with whatever index, as *some quantity.*

62. If any other term in the foregoing series, than the one just used, be selected, dividing it and all succeeding terms by h with whatever exponent h may have, the truth of the theorem becomes apparent. If the selected term be negative, it is shown to be less than the algebraic sum of all succeeding ones, because that sum approximates zero indefinitely with h, while the selected term cannot become greater. However, in the application of the theorem, we may change the signs of every term of the series, as we may wish to compare quantities simply in their amount of difference from zero.

The principle of the theorem is true with greater generality than the simple form $(x \pm h)^n$, for any function of $x \pm h$ developable in a similar series, may replace that of particular powers of $x \pm h$.

63. Although in a later section technical definitions of the terms *maximum* and *minimum* will be given, a general idea of these values of a function has already been given; we proceed now to determine, by the aid of the theorem, some of these values in problems, where a development gives rise to protracted series, and by the use of inequations.

PROBLEMS.

64. 1. Let it be required to determine whether the remainder expressed thus,

$$a x - x^3,$$

has a maximum or greatest value.

By inspection, the term having the highest power of x is observed to be negative, hence the remainder may evidently be infinite and negative; and, the terms being two, a maximum is probable.

Supposing x in the expression or function of x, to be such in value as to render the function of the greatest value, and it be suggested that x then receive the addition of a small quantity h, then the amounts appended to the respective terms are related thus:

$$a h < 3 x^2 h + 3 x h^2 + h^3;$$

i. e., $$a < 3 x^2 + (3 x + h) h.$$

Again, if h be subtracted from x in each term, the amounts subtracted from each term, treated here as *positive* for a *mutual* comparison only, are related thus:

$$a h > 3 x^2 h - 3 x h^2 + h^3;$$

i. e., $$a > 3 x^2 - (3 x - h) h,$$

where if $h = 0$, $\quad a = 3 x^2$,

and $$x = \pm \left(\frac{a}{3}\right)^{\frac{1}{2}}.$$

Here we find two answers, so we may infer that we ought to have asked if $a x - x^3$ may not also have an infinitely great *positive* value, which we should find to be true in the negative values of x. When the function is adapted to express its values with x negative, it is

$$(+ a \times - x) = - (- x \times - x \times - x),$$

that is, $$- a x + x^3, \text{ or } x^3 - a x.$$

3 *

In reasoning upon this basis, our signs of inequality would have been the reverse of what they have been, but the result the same.

2. Let the function of x be $a x^3 - x^4$.

On inspection, the function having different powers and the highest power of x in a negative term, it is probable there may be a maximum; considering x to be such that the function is at a maximum, and using h as before, we have,

$$3 a x^2 + (3 a x + a h) h < 4 x^3 + (6 x^2 h + 4 x h^2 + h^3) h;$$

and on diminution of x by the amount h,

$$3 a x^2 - (3 a x - a h) h > 4 x^3 - (6 x^2 h - 4 x h^2 + h^3) h;$$

where we have changed all the signs for the reason before given, and restored those within the parenthesis, because they are prefixed by *minus*.

These inequations must remain such after we have diminished the smaller member by dismissing an added term $(3 a x + a h) h$, and have increased a larger member by dismissing a subtractive quantity $(3 a x - a h) h$. So that representing the quantity within the parenthesis of the right hand by S and S', we have

$$3 a x^2 < 4 x^3 + S h,$$

$$3 a x^2 > 4 x^3 - S' h;$$

hence $$3 a x^2 = 4 x^3 \therefore x = \tfrac{3}{4} a.$$

There is no minimum value for $a x^3 - x^4$, because when it is regarded with respect to the negative values of x it appears as

$$- a x^3 - x^4;$$

for the function, as first given, is

$$ax^3 - x^4;$$

i.e., $a(+x\times+x\times+x)-(+x\times+x\times+x\times+x)$.

Now the sign of x changed gives us

$$-ax^3 - x^4,$$

or, $a(-x\times-x\times-x)-(-x\times-x\times-x\times-x)$.

3. To determine if the function bx^3+x^4 has a minimum or least value, at negative values of x.

4. To determine if the function $3x^3+7x-5x^2$ has a maximum or minimum value, or both.

Supposing x to take an increment h, and, whether we regard the maximum or minimum, we find, after coupling terms of like signs as distinct members of an inequation, and eliminating as before,

$$9x^2+7=10x;$$

which gives two values of x, one adapted to a maximum, the other to a minimum.

5. To determine the maximum or minimum value of the function:

$$\frac{5+x}{9+x}-8x.$$

Here in positive values of x the *minuend* cannot be greater than 1, nor less than $\frac{5}{9}$; but the subtrahend may be infinitely negative. So, regarding the *remainder* the greatest possible, we must have, if x be increased by h, and decreased by h:

$$\frac{5+x+h}{9+x+h}-\frac{5+x}{9+x}<8h;$$

$$\frac{5+x-h}{9+x-h}-\frac{5+x-h}{9+x-h}>8h;$$

reducing to a common denominator, *adding* terms and dividing by h, we have

$$\frac{9-5}{(9+x+h)(9+x)} < 8,$$

and
$$\frac{9-5}{(9+x-h)(9+x)} > 8;$$

whence, when $h = 0$, we have

$$\frac{9-5}{(9+x)^2} = 8;$$

which determines x, one of the two values of which are for a minimum.

6. Has $-x^2$ a maximum?

This is equivalent to $0 - x^2$; then as before,

$$0 < 2xh + h^2,$$

$$0 > 2xh - h^2,$$

$$\therefore x = 0 \text{ at the maximum.}$$

7. Required the value of x, which renders $x^4 + ax$ a maximum or minimum.

It may, by inspection, be inferred to be infinitely great, at one value of x, therefore it is probable there is only a minimum; using zero to represent a subtrahend, we have from

$$x^4 + ax - 0,$$

$$4x^3 + hS + a > 0,$$

and
$$4x^3 - hS' + a < 0;$$

$$\therefore 4x^3 + a = 0,$$

$$\therefore x = -\left(\frac{a}{4}\right)^{\frac{1}{3}},$$

and the function is at a minimum.

8. Has y a minimum in

$$y = x^3 + a x?$$

9. To determine whether y, the following function of x, has a maximum, and at what value of x:

$$y = \frac{3 + x^3}{1 + x^2}; \text{ i.e., } \frac{x^3}{1 + x^2} + \frac{3}{1 + x^2}$$

Here we have no visible expression of a negative term, but yet, when x is at some greatness, or greater than 1, the first of the resolved terms on growth of x tends to increase, and the second always to decrease while $x > 0$; but when x in value lies between zero and $\sqrt{3}$ the study of it will show that the reverse of the above is true of the first term, therefore there is a probability of a minimum. Hence, when the function is at a minimum, the term about to increase must do that faster than the other diminishes, so that

$$\frac{x^3 + 3x^2h + 3xh^2 + h^3}{1 + x^2 + 2xh + h^2} - \frac{x^3}{1 + x^2} > \frac{3}{1 + x^2} - \frac{3}{1 + x^2 + 2xh + h^2}$$

$$-\frac{x^3 - 3x^2h + 3xh^2 - h^3}{1 + x^2 - 2xh + h^2} + \frac{x^3}{1 + x^2} < -\frac{3}{1 + x^2} + \frac{3}{1 + x^2 - 2xh + h^2}$$

$$\therefore \frac{3x^2h + 3xh^2 + h^3 + x^2h^3 + x^4h}{(1 + x^2 + 2xh + h^2)(1 + x^2)} > \frac{6xh + 3h^2}{(1 + x^2 + 2xh + h^2)(1 + x^2)}$$

$$\frac{3x^2h - 3xh^2 + h^3 + x^2h^3 + x^4h}{(1 + x^2 - 2xh + h^2)(1 + x^2)} < \frac{6xh - 3h^2}{(1 + x^2 - 2xh + h^2)(1 + x^2)}.$$

Now disregarding denominators, because they are *common*, dividing by h, and using S and S' as before, we have,

$$3x^2 + x^4 + Sh > 6x + 3h,$$

$$3x^2 + x^4 - S'h < 6x - 3h,$$

$$\therefore 3x^2 + x^4 = 6x, \therefore x > 1 \text{ and } x < 2.$$

When $h = 0$, the denominators become $(1 + x^2)^2$.

65. These illustrations have been given in full, to show what course might be adopted for the solution of many problems. The following sections will show how the creation of the terms unused, may be dispensed with, by the use of *differentiation.*

Another advantage will be found to be the needlessness of any presumption in advance about maxima or minima as was apparently necessary above. Further, the successive maxima and minima of the same function will be beautifully deduced by the principles of differentiation.

These illustrations have also been extended to this degree, because the deductions are purely algebraic, pointing out forcibly the necessity of differentiation.

66. Thus far we have struggled for want of the language of expression, which the principles of differentiation are about to supply us, and the mode adopted, of solution by inequations, will be left as of no further use, and as likely to be encumbered with insuperable difficulties.

Such are the dilemmas into which those who slight the calculus must find themselves involved, who would nevertheless pursue its subjects of investigation, without its technical methods and language of expression.

SECTION VI.

DIFFERENTIAL OF A VARIABLE.—DIFFERENTIAL OF A FUNCTION.

67. It may have been seen in the solution of some problems that have been presented that a certain use was made, of the first one of those appended terms of increase or decrease, or rather in some cases, first set of terms including in one all those sub-terms which together are a factor to h, in composing one term of the function expanded to express its new consecutive value — that first term appended to the function in its primitive state.

Thus if the function of x were

$$x^3 + 5x^2 - 3x,$$

the set of sub-terms that compose *one* as recognized in the language of the Binomial Theorem, is

$$3x^2h + 10xh - 3h,$$

and the *one term* is

$$(3x^2 + 10x - 3)h;$$

the use referred to, consisted in equating such term with 0. This made an hypothesis for an inferred value of x, and an inferred characteristic for the function, while h became eliminated when made equal to 0.

68. Such an expression is the *differential* of the *function*, while h, when put at its limit, zero, in value, and called dx, is the *differential* of the *variable*, where d is a symbol, not a quantity, and never has an isolated position.

The differential of the function would then need to be symbolized as $d\ (Fx)$, or if y were the function's representative in amount, we may put dy for $d\ (Fx)$. The expression Fx means a function of x, hence F is not separable as a factor.

69. The expression $d x$ has one advantage of zero or 0, that it designates a relation to x, in contradistinction from another variable z, or w, which may be associated in the same expression; and from $d y$.

70. The act of taking the differential of a function is called *Differentiation.*

71. When the function we may wish to differentiate, is of one term, not fractional, and has a power of which the variable is the root, and the index is a whole positive number, the Binomial Theorem teaches at once that:

72. *We should multiply together the index of the power, the power having its index diminished by one, the constant factor if there be such, and the differential of the variable. The product is the differential required.*

PROBLEMS.

73. 1. What is the differential of the function x^2 of the variable x, the function being equated with y.

Ans. $d y = 2 x d x$.

2. Required the differential of $b x^3 = y$.

Ans. $d y = 3 b x^2 d x$.

3. Required the differential of $7 c x^4 = y$.

Ans. $d y = 28 c x^3 d x$.

4. Required the differential of $150 x^n = y$.

Ans. $d y = 150 n x^{n-1} d x$.

5. Required the differential of $100 x = y$.

Ans. $d y = 100 d x$.

It may be readily inferred that if the function has a term or terms, in which the variable does not enter, such terms are constant, and contribute nothing to the differential. Such term, however, affects the function by supplying a basis of value common to every value of it.

6. Required the differential of $10010\,x^2 + 5 = y$.

Ans. $d\,y = 20020\,x\,d\,x$.

7. Required the differential of $a - 5\,x^2 = y$.

Ans. $d\,y = -10\,x\,d\,x$.

74. If we attend to the origin of the numerical coefficients of the second term of the Binomial series (the units of which are a transfer of those of the index), as well as to the origin of such index, we perceive that

$$x^2 = x\,.\,x' \text{ and } d\,(x^2) = 2\,x\,d\,x = x\,d\,x' + x'\,d\,x,$$

$$x^3 = x\,.\,x'\,.\,x'' \text{ and } d\,(x^3) = 3\,x^2\,d\,x =$$
$$x'\,.\,x''\,d\,x + x\,.\,x''\,d\,x' + x\,.\,x'\,d\,x'',$$

where we have placed accents, to denote the source of the different elements, and the mode by which they would contribute to the result, if the quantities x, x', and x'' were not alike. The condensation takes place because they are presumed to be alike.

8. Let it be required to express in detail as above, the differential of x^4 or $x\,.\,x'\,.\,x''\,.\,x'''$.

9. How then must the differential of such a function of x as $y = x\,(x - a)$ be taken if we do not choose to perform the multiplication?

Ans. $d\,y = (x - a)\,d\,x + x\,d\,(x - a)$.

10. Discover the identity of this answer with the differential of $x^2 - a\,x$, which is the above function after multiplication has been performed.

75. It is then sufficiently evident that to differentiate a $F\,x$, which is a product of several functions of the same variable:

We should multiply the differential of each factor by the product of the other factors and add the obtained products — remembering the algebraic meaning of the word *add.*

4

1. Required dy from $(a-x^2)(x-1)=y$.

Ans. $dy=$

2. Required dy from $(x^3+a)(3x^2+b)=y$.

Ans. $dy=(15x^4+3bx^2+6ax)\,dx$.

3. Required dy from $x^2(x-a)^6=y$.

Ans. $dy=2x(x-a)^6\,dx+6x^2(x-a)^5\,dx$.

In differentiating the function $5x$, and finding it to be $5\,dx$, we may observe that $5x=x+x+x+x+x$, and $5\,dx=dx+dx+dx+dx+dx$.

It is almost self-evident that if we have such a function of x as

$$ax^2+x-x^3=y,$$

and x take an increment dx, y is affected by the several differentials of the terms which compose y. Hence:

76. To differentiate the one function of a variable, which is a sum or difference of certain *other* functions of it; we must *take the differentials of the component functions and connect them* by the signs by which they affect the one function.

4. Hence the dy of $ax^2+x-x^3=y$ is

$$dy=(2ax+1-3x^2)\,dx.$$

77. To differentiate a function of a variable when it is of a fractional form.

5. Let $$\frac{x^2}{(x-a)^3}=y.$$

For convenience replace the numerator by N and the denominator by D, then

$$y=\frac{N}{D},$$

$$\therefore D\times y=N;$$

where Dy is a product of two quantities,

$$\therefore d\,(D\,y) = d\,N,$$

i. e.,
$$y\,d\,D + D\,d\,y = d\,N,$$

$$\therefore d\,y = \frac{d\,N - y\,d\,D}{D},$$

replacing the value of y

$$d\,y = \frac{D\,d\,N - N\,d\,D}{D^2}.$$

78. Wherefore, the rule for differentiating a fraction is found to be:

From the Denominator multiplied by the differential of the Numerator, subtract the Numerator multiplied by the differential of the Denominator, and divide the remainder by the second power of the Denominator.

Or, by a discovery of an artificial device for aiding the memory, and simulating the form:

Fr*om* de*nom*' by differ'-of nu*mer*' —
subt*ra't* nume*rat*' by differen*tia*' of denom*ina*',

divide the remainder by (denominator) ².

6. Required the differential of $y = \frac{a\,x - b}{c\,x^2}$.

Ans. $d\,y = \frac{a\,c\,x\,d\,x - 2\,(a\,x - b)\,c\,d\,x}{c^2\,x^3}$.

79. If the variable does not appear in the denominator of the function given, it is evident that such denominator, by taking unity as its own separable numerator, may be separated from the function as a constant factor. The rule results in this.

7. What is $d\,y$ in the case of $y = \frac{x^2}{a} = \frac{1}{a} \times x^2$?

Ans. $d\,y = \frac{2}{a}\,x\,d\,x$.

8. Required the dy from $y=x^{-3}$; i. e., $\frac{1}{x^3}$.

$$d\left(\frac{1}{x^3}\right)=-\frac{3x^2}{x^6}=-3x^{-4}dx,$$

which is in accordance with the rule for positive indexes.

80. Let it be required to differentiate a function of a variable, which is at once a power with some root of that power expressed:

9. Let $y=x^{\frac{2}{3}}$,
then $y^3=x^2$,

$$\therefore 3y^2dy=2x\,dx,$$

$$dy=\frac{2x\,dx}{3y^2}=\frac{2x\,dx}{3x^{\frac{4}{3}}}=\frac{2}{3}x^{-\frac{1}{3}}dx.$$

10. Next let

$$y=x^{-\frac{2}{3}}=\frac{1}{x^{\frac{2}{3}}},$$

$$dy=-\frac{\frac{2}{3}x^{-\frac{1}{3}}}{x^{\frac{4}{3}}}dx=-\frac{2}{3}x^{-\frac{5}{3}}dx.$$

11. Differentiate $x^{\frac{m}{n}}=y$, m and n being any whole numbers, and the fraction $\frac{m}{n}$, positive or negative.

$$dy=\frac{m}{n}x^{\frac{m}{n}-1}dx.$$

12. Differentiate $x^{-\frac{m}{n}}=y$.

$$dy=-\frac{m}{n}x^{-\frac{m}{n}-1}dx.$$

81. Hence, generally to differentiate a function of a variable, expressed by the variable at a power denoted by an index, positive or negative, whole or fractional, the power having, or not having, a constant as factor:

Bring down that index, with its sign, to be a coefficient (or a part of it); annex any constant which was an original factor; then the variable, having now its index diminished by unity; then the differential of the variable;—their continued product is the differential required.

82. But if the function given, and having such exponent, is not the variable, but some function of it, the above rule still holds; using, instead of the word *variable*, that function of it which is the root, for:

13. What is dy in $(a-x^2)^2=y$?

$$\text{Using } u \text{ for } a-x^2,\ du=-2x\,dx,$$
$$\therefore dy=2u\,du=-4ux\,dx,$$
$$dy=-4x(a-x^2)\,dx.$$

By the above rule the differential of any root is obtained.

14. Required dy from $y=\sqrt{x}=x^{\frac{1}{2}}$.

$$\text{Ans. } dy=d(x^{\frac{1}{2}})=\pm\frac{dx}{2\sqrt{x}}.$$

83. From the foregoing we infer, that if roots of powers, or powers of roots, are expressed by fractional or other indexes, their differentiation is made plain.

84. But a root affected with such index may be some function of the variable.

15. Required the differential of $y=(a-x^2)^{\frac{1}{2}}$.

$$\text{Putting } u=a-x^2,$$
$$\therefore du=-2x\,dx,$$
$$y=u^{\frac{1}{2}},$$
$$\therefore dy=d(u^{\frac{1}{2}})=\frac{du}{2u^{-\frac{1}{2}}}=-\frac{x\,dx}{(a-x^2)^{-\frac{1}{2}}}.$$

16. Required the differential of the function of x, viz.:

$$y = \frac{a + x}{\sqrt{a - x}}.$$

Ans. $d\,y = \pm \dfrac{3\,a - x}{2\,(a - x)^{\frac{3}{2}}}\,d\,x.$

17. Required the differential of

$$y = \frac{b\,x + x^{\frac{1}{2}}}{(a - b\,x^3)^{\frac{1}{2}}}.$$

SECTION VII.

FIRST DIFFERENTIAL COEFFICIENT.

85. Whenever a differential of a function is obtained, it will have been seen that the differential of the variable ($d\,x$), is always found to be a factor of it. If now we divide by this $d\,x$, its natural algebraic position is as a denominator to $d\,y$; thus, if we have

$$y = F\,x,$$

$$\therefore d\,y = d\,(F\,x),$$

$$\therefore \frac{d\,y}{d\,x} = d\,(F\,x) \div d\,x;$$

and $\dfrac{d\,y}{d\,x}$ will be an expression in the *general* form for the *differential coefficient.* The *actual* quantity for the differential coefficient, in respect to a *particular function,* will be that with which $\dfrac{d\,y}{d\,x}$ is equated.

86. *First Differential Coefficient* is the nomenclature of Leibnitz. *First Derivative* or *First Derived Function* is by that of Lagrange. If the word *first* is omitted, and second or third, etc., is not expressed, the first is understood. We use the nomenclature of Leibnitz, but mention that of Lagrange as in use.

87. Since the value of a ratio is properly expressed as a quotient and by a fraction, $\frac{dy}{dx}$ (meaning always by this some special quantity *derived* from an actual function) is the value of the ratio, between the amount of the change of the function and the corresponding change of the variable; more strictly it is the limit of that ratio, as it becomes when the increment of the variable h is made zero, when we call it dx.

The dif. coef. of $a x^3 = y$, is $3 a x^2 = \frac{dy}{dx}$.

$$\therefore dy : dx :: 3 a x^2 : 1;$$

that is, $\qquad 0 : 0 :: 0 \times 3 a x^2 : 0 \times 1.$

88. The expression for a first differential coefficient, in its general form, is $\frac{dy}{dx}$. In a particular function of one variable, the y in dy is that particular function, and the x in dx is the particular variable. The expresssion $\frac{dy}{dx}$ is, in respect to its signification for a value, the same as $\frac{0}{0}$ (and this we have already found in algebra may have any value whatsoever), with this advantage over $\frac{0}{0}$, that it preserves a reference to its origin, as a means of determining its value.

89. The value of a fraction of which both the numerator and denominator is 0, is determined by the disposition of those quantities each of which may have actually become 0, to emerge from that state; then they come into being

with a ratio, the value of which is determined by their attendant factors:

thus, $$\frac{0 \times a}{0 \times b} = \frac{0}{0} = \frac{a}{b};$$

or rather, $$\frac{(c - x)\,a}{(c - x)\,b} = \frac{a}{b},$$

which we deduce when x has any value, inclusive of $x = c$.

90. Let it be required to find the first differential coefficient of the following functions of a single variable x, each standing equated with y as the same value with each function; but severally or disconnectedly, the several functions being absolutely independent of each other.

1. Given $y = 152783\,x$.

$$\therefore \frac{d\,y}{d\,x} = 152783.$$

2. Given $y = \frac{a}{150000}\,x^3$.

$$\therefore \frac{d\,y}{d\,x} = \frac{a\,x^2}{50000}.$$

3. Given $y = (b\,x + a)\,x^5$.

$$\therefore \frac{d\,y}{d\,x} = (6\,b\,x + 5\,a)\,x^4.$$

4. Given $y = (17\,a\,b - 1)\,x + 34\,a\,x^2$.

$$\therefore \frac{d\,y}{d\,x} = 17\,a\,b - 1 + 68\,a\,x.$$

5. Given $y = a\,b\,x^5 - (a\,x + b\,x^2)^5$.

$$\therefore \frac{d\,y}{d\,x} = 5\,a\,b\,x^4 - 5\,(a\,x + b\,x^2)^4\,(a + 2\,b\,x).$$

6. Given $y = (a x + b x^2 - c)^2$.

$$\therefore \frac{dy}{dx} = 2 (a x + b x^2 - c)(a + 2 b x).$$

7. Given $y = \sqrt{a x + c x^3}$.

$$dy = \frac{(a + 3 c x^2) \, dx}{2 \sqrt{a x + c x^3}}, \; \frac{dy}{dx} = \frac{a + 3 c x^2}{2 \sqrt{a x + c x^3}}.$$

8. Given $y = (a \sqrt{x} + x^2)^2 + x^5 - (c x^n - 1)^3$.

$$\frac{dy}{dx} =$$

$$2 (a \sqrt{x} + x^2) \left(\frac{a}{2 \sqrt{x}} + 2 x \right) + 5 x^4 - 3^n (c x^n - 1)^2 c x^{n-1}.$$

9. Given $y = (a + b x) x^3$.

$$\frac{dy}{dx} = b x^3 + 3 x^2 (a + b x).$$

10. Given $y = \frac{x^2}{x + 1}$.

$$\frac{dy}{dx} = \frac{x^2 + 2 x}{(x + 1)^2}.$$

11. Given $y = \frac{a x^2}{b + x}$.

$$\frac{dy}{dx} = \frac{(b + x) 2 a x - a x^2}{b^2 + 2 b x + x^2}.$$

12. Given $y = - \frac{c}{2 x^2}$.

$$\frac{dy}{dx} = c x^{-3}.$$

13. Given $y = \frac{x}{\sqrt{x - a}}$.

$$\frac{dy}{dx} = \pm \frac{2 (x - a)^{\frac{1}{2}} - x (x - a)^{-\frac{1}{2}}}{2 x - 2 a}.$$

14. What is the first differential coefficient respectively of the following (functions of x) $= y$?

$$(b + a\,x^2)^{\frac{5}{4}} = y.$$

$$\text{Ans. } \frac{d\,y}{d\,x} = \frac{5\,a\,x}{2}\sqrt[4]{b + a\,x^2}.$$

15. $$a + \frac{4\sqrt{x}}{3 + x^2} = y.$$

$$\text{Ans. } \frac{d\,y}{d\,x} = \frac{6\,(1 - x^2)}{(3 + x^2)^2\sqrt{x}}.$$

16. $$\frac{x}{x + \sqrt{1 - x^2}} = y.$$

$$\text{Ans. } \frac{d\,y}{d\,x} = \frac{1}{\sqrt{1 - x^2}\,(1 + 2\,x\sqrt{1 - x^2})}.$$

17. $$(x + a)\sqrt{4\,b^2 - (x - a)^2} = y.$$

$$\text{Ans. } \frac{d\,y}{d\,x} = (x + a)\frac{a - x}{\sqrt{4\,b^2 - (x - a)^2}} + \sqrt{4\,b^2 - (x - a)^2}.$$

18. $$a + b\sqrt{x} - \frac{c}{x} = y.$$

$$\text{Ans. } \frac{d\,y}{d\,x} = \frac{b}{2\sqrt{x}} + \frac{c}{x^2}.$$

19. $$(a\,x^3 + b)^2 + 2\sqrt{a^2 - x^2}\,(x - b) = y.$$

$$\text{Ans. } \frac{d\,y}{d\,x} = 6\,a\,x^2\,(a\,x^3 + b) + \frac{2\,(a^2 - 2\,x^2 + b\,x)}{\sqrt{a^2 - x^2}}.$$

91. The reciprocal of a differential coefficient of a $F\,x = y$ being evidently $\frac{d\,x}{d\,.\,y}$ in general expression, it may be found with great facility in a particular case, — it being necessary, simply, to make the quantity equated with $\frac{dy}{dx}$ a denominator of unity; or, if the dif. coef. is already in the form of a fraction, to exchange numerator for denominator. Hence, whatever purpose $\frac{dy}{dx}$ may serve in reference to $Fx = y$, where x is the independent variable, and the value of the function, i. e., y is the dependent variable, is subserved by $\frac{dx}{dy}$ in the case in which y is to be regarded as the independent variable of the function whose value is x, the dependent variable.

20. Required $\frac{dx}{dy}$ in

$$c + a\sqrt{x} - \frac{b}{x} = y.$$

Ans. $\frac{2x^2}{ax^{\frac{3}{2}} + 2b}$.

SECTION VIII.

USE OF FIRST DIFFERENTIAL COEFFICIENTS.

92. We have shown how a first differential coefficient of any explicit function of a single variable may be obtained, and have deduced a general expression for it, that is, $\frac{dy}{dx}$, which is consistent with y assumed as the amount of the value of the function, with x as the variable, and with each of their differentials, when made $= 0$.

The particular differential coefficient is that quantity with which $\frac{dy}{dx}$ is found to be equated.

When a first differential coefficient contains the variable x, it has a value, when a value is assumed for x. Or, a value may be assumed for the dif. coef., within the limits of possible values, which may be equated with the actual particular dif. coef. and the corresponding value of x deduced.

This same value of x referred back to the function, and x replaced by it, gives us also the corresponding value of the function, with which such value of the dif. coef. agrees.

Wherever we mention the singular meaning of the word *value*, of a variable or a function, let the plural, *or values*, be understood in the same connection.

93. When a first dif. coef. consists of several terms (which we have called *sub-terms* in their relation to the binomial theorem), connected by the diverse signs $+$ and $-$, it must be evident that the resultant sign for the dif. coef. depends on the value of that variable; which, indeed, is true in determining the value of a function.

94. As we have seen, a first dif. coef. may have only one possible value, i. e., when it does not contain the variable. It may also have every conceivable value from $-\infty$ to $+\infty$.

95. We shall use, as will be seen, the differentiation *of a dif. coef.* for its service, in investigating a dif. coef. as a *derived function.*

96. Sometimes, in mathematical investigations, we first arrive at a quantity, which it is necessary to regard as a differential, or a dif. coef., in a case when the future object of search is for its function which we have not possessed; this search, when generalized for every possible case, may be very elaborate, and is denominated the Integral Calculus.

97. We ought, in a previous section, to have observed, in the study of functions in connection with their variables, that we have four conditions to regard with respect to positive and negative values.

1. The *function* may be *positive* in value, while the *variable* is *positive*, thus: $+ a \times + x = a x$.

2. The *function* may be *positive* in value while the *variable* is *negative*, thus: $(- x)^2 = x^2$ or $- a \times - x = a x$.

3. The *function* may be *negative* in value while the *variable* is *positive*, thus: $(- a \times + x) = - a x$, or $- (+ x) = - x$.

4. The *function* may be *negative* in value while the *variable* is *negative*, thus: $(a \times - x) = - a x$.

If, under the 3d head, we should have given as an illustration $- x^3 = - (+ x . + x . + x)$, we remark that the sign $-$ is the sign that indicates how the amount x^3 is to be applied as a value for the function; it is not the sign of the individual x.

Since x alone is by no definition excluded from being a function of x, it seems singular that $- x$ as a function contains $+ x$ nevertheless as a variable. But in such use the negative sign is applied to x only for its value as a function. Thus, if $- x$ be the function which is also y, we have,

$$y = - x,$$

that is, $$- y = + x;$$

so that the indication of $- x$ as a function is, that the function has a negative value to the extent of $+ x$.

It is then superfluous to remark, that it becomes unnecessary to use such language as, a function of $- x$, or $F(- x)$, but rather, a function of $+ x$, $F(+ x)$; i. e., $F x$

with reference sometimes, as it may be, to negative values of x in it.

Problems.

1. Required the value of x when it is increasing $\frac{1}{45}$ as fast as $27x + 3x^2$. Ans. 3.

2. Required the value of x when it is increasing 1000000 times as fast as $27x + 3x^2$. Ans. $-4\frac{2999999}{6000000}$.

A quantity is considered to be increasing when it is becoming a smaller negative one.

3. To determine the value of x when it is one and the same variable in x^2 and x^3, at the values of these functions when they are increasing with equal rapidity. Calling $x^2 = y$ and $x^3 = y'$, we have,

$$\frac{dy}{dx} = \frac{dy'}{dx} = 2x = 3x^2,$$

$$\therefore x = \pm \frac{2}{3}.$$

4. Suppose the Boston and Maine Railroad running north from Boston, and the Old Colony and Fall River Railroad running south from Boston, to be one continuous north and south railroad passing through Boston, and Worcester to be 40 miles west of Boston. The cars on one of these roads, being 30 miles from Boston, are running south at 23 miles an hour; how fast are they affecting their distance from Worcester?

Let $x =$ the distance of the cars from Boston in miles.

$\therefore (40^2 + x^2)^{\frac{1}{2}}$ is the distance from Worcester, which call y.

$$\therefore \frac{dy}{dx} = \pm \frac{x}{(40^2 + x^2)^{\frac{1}{2}}} = \pm \frac{3}{5},$$

$$x \text{ being} = 30.$$

Now, if x vary 23 times the natural x as related to y, y will vary $23 \times \pm \frac{3}{5} = \pm 13\frac{4}{5}$.

Ans. $\pm 13\frac{4}{5}$ miles. The positive answer being adapted to increase of distance from W., the cars being south of B.

The ambiguous answer is adapted to each of the railroads, the direction of the motion being the same.

5. At the point where the cars, running north 32 miles per hour, are affecting their distance from Worcester 18 miles per hour, how far are they from Boston?

Here, $$\frac{dy}{dx} = \frac{32x}{(40^2 + x^2)^{\frac{1}{2}}} = 18,$$

and $$x = 22.7 \text{ miles}.$$

6. A farmer is raising 1000 swine, which, on their attaining a certain weight, estimated here by their pork, he intends to slaughter and pack in casks, in connection with 2000 pounds of other pork, which is already stored in waiting; the cooper, without much consideration, had contracted to make the casks, each so as to contain one animal and 60 pounds more; but the farmer has not decided at what weight of growth to slaughter them. How heavy is one when, allowing it afterwards to grow, the growth will be at the rate of one pound, while the number of casks required changes one cask? and will it be increase or decrease of the number of casks?

Ans. Between 180 and 181 pounds; casks to increase.

7. Examine y, y', and y'' in these three functions of x, and determine which never increases and which never decreases at positive values of x; and *vice versa* with negative values of x; and which never changes at either positive or negative values of x, and simplify the form of one of them, and possibly eliminate x.

8. $$y = \frac{33x + 2000}{x + 60}.$$

9. $$y' = \frac{33\frac{1}{3}x + 2000}{x + 60}.$$

10. $$y'' = \frac{34x + 2000}{x + 60}.$$

11. Does one of the above expressions, and which, fail to be a function of x, in such sense that the change of x produces any effect on the value of the expression, which is to fail entirely?

12. A man having 4320 dollars, purchased a horse, and with the remainder of his money purchased sheep, at such rate, that the money expended for a horse would buy as many sheep as they were worth dollars apiece. How, with the preservation of this relation, would the number of sheep purchased, tend to vary, compared with the number of dollars paid for the horse, if he pays for him 36 dollars, while in the contemplation of paying a greater sum than that?

Ans. The number of sheep will diminish $10\frac{8}{100}$ times as much of a sheep as of a dollar more that should be paid for a horse, or at the rate of $10\frac{8}{100}$ sheep for one dollar additional paid for a horse.

When a coefficient has the ambiguous sign, we are often able, as here, to prefer one to the other from positive knowledge. The cost of a sheep is not to be supposed a negative sum of dollars.

13. A boy was holding by a cord in his hand a horse, the cord attached at the horse's mouth and held upon the level ground, thus permitting the horse to eat the grass upon a circle of ground; but having given the animal 90 feet, he commences drawing the cord in at the rate of 3 feet per second. How many square feet per second was the circular plot of grass diminishing? that plot which the conditions allow the animal, if he were sufficiently active to avail himself of it.

14. The boy begins to climb a tree 5 inches per second, and to let out cord 4 inches per second. How fast per second is the horse's circle changing in square feet, when the boy gets 10 feet high and has let out 70 feet of cord?

15. While climbing as above, and when 10 feet high, how fast should he pay out cord that the circle may remain stationary?

16. A ship is sailing north-west at 15 miles an hour; at what rate is she gaining in north latitude per hour?

Ans. 10.601 miles.

17. On the discharging of coal from a vessel, it is raised high in the air and thrown down into heaps, which may be cones. On one occasion, the height of the heap was always $\frac{2}{5}$ of the diameter of the base; when its height was 15 feet, and the solid contents were increasing at the rate of 160 cubic feet per hour, how fast per hour was the height increasing?

18. How fast per hour was the height increasing when it had become 32 feet high, the conic heap gaining 160 cubic feet per hour, and being always a *similar* heap?

19. A boy, amusing himself by throwing stones into a pond of still water, to see the circle of waves expand, perceived that the diameter of one circle increased 3 feet per second when it had become 24 feet in diameter; how fast was the area increasing in square feet at that instant?

20. It is required to divide 70 into two such parts, that the suggestion of increasing one part afterwards at the expense of the other, would implicate an increase of their product 40 times as fast as that one part should increase.

Ans. 15 and 55; the 15 to increase.

21. At a pin factory, a certain number of pins are stuck in a row in the papers; 3 more than that number of rows are put in a paper, and one less than that same number of papers are put in a package. It being suggested to diminish the number of pins in a row, how does that qualify

the change of the number in a package, while there happens to be 16 in a row?

There is this disability in this problem: the idea of fractional pin cannot be entertained, and the change of an entire pin cuts off the application of the principle of the initial ratio. Those quantities, a part of which can be practically considered as of the same nature as the whole, are best adapted to investigation and study by the calculus.

The disability is not mathematical, but concerns that practical economy by which events and things are submitted to calculation.

22. Some boys rolling a spherical ball of snow, observed that when it was 28 inches in diameter, it was increasing in diameter 5 inches per minute; how many cubic inches per minute were its solid contents increasing?

23. A certain cook adopted 200 handfuls and 300 ounces of flour, to be made into cakes, of $\frac{1}{8}$ of such handful each; if this would make 1840 cakes, what disposition had the number of cakes to vary, if the number of ounces in a handful were diminished the least amount.

Ans. Cakes increase in number at the rate of 24 to the first ounce that should be diminished from the handful, the weight of the cake of course diminishing.

24. A glazier prepared a quantity of putty sufficient to set 100 panes of glass, with an excess of 180 ounces; but at this stage, receiving orders to set glass of such size of pane as to require per pane more putty than his first estimate, it is required to determine how the number of panes, which he may be able to set with the whole of the putty, will vary with the increase of the number of ounces requisite for one pane: 1st, to determine this by a general expression; and 2d, what the amount of the ratio is, in the

particular case, when he was to use 3 ounces in setting his first panes.

2d Ans. Panes diminish 20 times as fast in number as the putty per pane increases per ounce, i. e., while 3 oz. suffice per pane in the first estimate.

25. Will y, or the sum of the following quotients, increase or decrease, when 7 is replaced by a numerical quantity just larger, y being a function of 7; 7 temporarily supplying the place of x?

$$\frac{25}{7}+\frac{7}{5}=y.$$

26. A person, thinking to propound a puzzle, said he was in the habit of purchasing the article A, at the rate of one dollar per ounce, and on each such occasion, of purchasing also the article B, paying the same amount of money for the whole of the article B as for the whole of A. Now, on every occasion of these two associated purchases, the weight of the amount purchased of A, added to that of B, made 20 ounces. But it is not intended that on all occasions the weights of A were the same, but that they varied indefinitely. On the occasion when he may have purchased $3\frac{1}{3}$ ounces of A, if we proceed to consider the occasion when he may have purchased any the least more of A, it is required to determine how we are to find the price per ounce of B to change between the corresponding occasions.

Ans. Number of ounces of A to increase $\frac{9}{125}$ as fast as the price in dollars of B per ounce.

27. It is required to determine the price per ounce of the article B, on an occasion when, on comparing the price with that of the same article on another occasion, in passing to which there may have been an increase, the least possible, in price, the corresponding increase in the amount of A purchased must have been equal to it; the

ounces of the A and the dollars per ounce of B being compared numerically in units. Ans. $3\frac{47}{100}$ dolls.

28. There are two numerical quantities, x and y, and they are such that 5000 times x plus 3 times y, are always equal to 200 times x times y. It is required to determine if x can have the value 1500, in which case it is required to determine the corresponding value of y.

Ans. $y = 25\frac{21}{299997}$.

29. It is required to determine how x and y, at the above values, are disposed to change their values.

Ans. y will be disposed to diminish $\frac{15000}{90000000009}$ as fast as x to increase.

SECTION IX.

SUCCESSIVE DIFFERENTIATION. — SECOND, THIRD, ETC., DIFFERENTIALS.

98. When a first differential coefficient contains the variable, it is evidently a function of it, called *first derived function.* When the variable x of the primitive function varies, the differential coefficient will itself vary. Hence those suppositions in some problems, where a differential coefficient contains x, and is supposed or inferred to have a value, give correct results only for an exactly specified, or exactly inferable, value of x, and of the function.

Where a certain rate has been inferred for the cars to affect their distance from Worcester, as at so many miles per hour, the execution of it could not take place during a minute or an entire rod. The supposition and the inference are good for only an instant of time, and at a point only in place; at the succeeding instant of place and time, the differential coefficient has changed, the distance from Worcester has changed, and the rate of the *changing* of that distance has changed. (Page 50, Prob. 4.)

What is more natural, then, than to employ the same means in determining the *general* character of the *derived* function, that we already partially have, in regard to the primitive—differentiate it? and why not, perhaps, continue to differentiate the *second derived* function, or second differential coefficient?

Let it be required to differentiate successively $y = 15x^3$, and while doing it, to evolve the notation for these acts; and for the results

we have $$y = 15x^3,$$

$$\therefore \frac{dy}{dx} = 45x^2,$$

i. e., $$\frac{1}{dx} \times dy = 45x^2,$$

where $\frac{1}{dx}$ is a factor in the notation $\frac{dy}{dx}$, and should be supposed constant, not only because we are able, in case of a fraction which is to change its value, to throw that change entirely into the numerator *or* denominator at will; but because we set out in this case with a function y, which would change if x did, and we suppose x to change by an amount which we can exactly define, h or dx, with the expectation of throwing all quantity that must have any other nature than an exact and certain one, which may be arbitrarily made uniform, upon the change of y. Hence, differentiating $45x^2$, and expressing that of its equal in the notation, we have

$$\frac{1}{dx} \times d(dy) = 90x\,dx.$$

We evidently can, and ought to, assume the *independent* variable to vary uniformly, thus allowing the results to be all made manifest in the value of the function which depends upon it.

If now we agree, as is the custom, to represent $d\,(d\,y)$ the differential of the first differential of y by $d^2\,y$, where the 2 does not signify a power, but the second act of differentiation, we shall have

$$\frac{1}{d\,x} \times d^2\,y = 90\,x\,d\,x \text{ and } \frac{d^2\,y}{d\,x^2} = 90\,x,$$

where $d\,x^2$ is by proper algebraic act the second power of $d\,x$, and is not the same as $d\,(x^2)$.

In a similar manner may *third*, *fourth*, and more differentiations be performed on a function that admits of them, and the notation $\frac{d^3\,y}{d\,x^3}$ and $\frac{d^4\,y}{d\,x^4}$, etc., be derived.

In the last function, viz., $15\,x^3$,

$$\frac{d^3\,y}{d\,x^3} = 90, \frac{d^4\,y}{d\,x^4} = 0, \text{etc.}$$

Hence, differentiations will terminate of functions in which the variable appears at a power denoted by a whole and positive index, and when the function is not of a fractional form, the variable in a denominator. In such cases differentiation may never terminate.

1. Differentiate, successively, $6\,x^3 - 5\,x^2 + 60\,x = y$.

$$\frac{d\,y}{d\,x} = 18\,x^2 - 10\,x + 60, \frac{d^2\,y}{d\,x^2} = 36\,x - 10;$$

i. e., $$\frac{1}{d\,x^2} \times d^2\,y = 36\,x - 10,$$

$$\frac{1}{d\,x^2} \times d\,(d^2\,y), \text{ i. e., } \frac{d^3\,y}{d\,x^2} = 36\,d\,x,$$

$$\therefore \frac{d^3\,y}{d\,x^3} = 36, \frac{d^4\,y}{d\,x^4} = 0.$$

2. Differentiate $(a - x)\sqrt{x} + \sqrt{y} = 0$, a $F(x, y) = 0$; i. e., a function of x and $y = 0$.

$$\frac{d^2 y}{d x^2} = 6 x - 4 a.$$

3. Given $4 u^2 + u = y$, in which u is this $F x$, viz., $u = 3 x^2$, to find $\frac{d^2 y}{d x^2}$.

4. Required $\frac{d^2 y}{d x^2}$ in $6 x^3 - 5 x^2 + 60 x + 10 = y$.

Ans. $36 x - 10$.

5. When the sub-terms are reduced to a resultant term, what is the sign of $36 x - 10$ when $x > 3$? When $x < \frac{1}{4}$?

6. Given $28 x^2 - y^2 = 0$, to find $F x = y$, $\frac{d y}{d x}$ and $\frac{d^2 y}{d x^2}$.

Ans. $y = x \sqrt{28}$, $\frac{d y}{d x} = \sqrt{28}$, $\frac{d^2 y}{d x^2} = 0$.

7. Given $x^2 + \frac{a^2}{b^2}(2 b x - x^2) = y$, to find $\frac{d^2 y}{d x^2}$.

Ans. $2\left(1 - \frac{a^2}{b^2}\right)$.

8. Given $24 x^2 - y^3 + 10 x = 0$, to find $\frac{d^3 y}{d x^3}$.

9. Given $x^6 - 2 x^3 y + y^2 - x^2 = 0$, to find $\frac{d^2 y}{d x^2}$.

Ans. $6 x$.

10. Given $4 x^2 + x = z$, and $3 z^2 + 2 z = y$, to find $\frac{d^2 y}{d x^2}$.

The following analogy is worthy of note: If we take the third powers of the natural numbers 1, 2, 3, etc., and then their differences, and then the differences of these differences, and so on, we have

$$
\begin{array}{ccccccccccccccc}
0 & & 1 & & 8 & & 27 & & 64 & & 125 & & 216 & & 343, \text{ etc.,} \\
& 1 & & 7 & & 19 & & 37 & & 61 & & 91 & & 127, \text{ etc.,} & \\
& & 6 & & 12 & & 18 & & 24 & & 30 & & 36, \text{ etc.,} & & \\
& & & 6 & & 6 & & 6 & & 6 & & 6, \text{ etc.,} & & & \\
& & & & 0 & & 0 & & 0 & & 0, \text{ etc.} & & & &
\end{array}
$$

If we take $y = x^3$ and differentiate, we have

$$\frac{d\,y}{d\,x} = 3\,x^2,$$

$$\frac{d^2\,y}{d\,x^2} = 6\,x,$$

$$\frac{d^3\,y}{d\,x^3} = 6,$$

$$\frac{d^4\,y}{d\,x^4} = 0.$$

If a first dif. coef. for any determining reason has the value 0, we may determine what the function is about to do, in regard to increase or decrease, by the sign of the second dif. coef., which is an important principle. If the second dif. coef. $= 0$, we determine it by the sign of the third, and so on.

The reason why a dif. coef. may have the value 0, is owing to an aggregate of sub-terms with different signs, which may compose it, and a *particular* value of the variable also supposed. Or if there be but one term, and it contain the variable as a factor, it $= 0$ when $x = 0$.

The process of successive differentiations may terminate, or never; some fractional forms are of the latter character, as well as those with negative and fractional indexes.

Differential coefficients may change their signs as the variable is traced through successive values. Of course the general form of notation cannot show this. Particular values of the variable determine this change of sign.

When a dif. coef. does not contain the variable, it may not have the value zero.

To determine whether a dif. coef. can have the value zero, we equate it with zero and reduce the equation.

SECTION X.

TAYLOR'S THEOREM.

99. It is the purpose of Taylor's Theorem to lay down, in a general expression, in the form of a series, when this is possible, the different orders of differential coefficients, with their signs and necessary factors attached, according to which (primarily) a function of a single variable is developed, with an increment or decrement to the variable, and in the order of the increase of the integral powers of the increment or decrement, by the natural series of numbers.

Let $y = Fx$ and $Y = F(x + h)$, and assume

$$Y = y + Ah + Bh^2 + Ch^3 +, \text{ etc.},$$

where A, B, C, etc., are quantities which x or constants can express, and which are now wanted to replace A, B, C, etc., with.

Remembering that $x+h$ is the new variable, and that we can ascribe a new variation of $x+h$ to either x or h at pleasure, while the other of the two will be constant, we have, by differentiating, first, with respect to h,

$$\frac{dY}{dh}=A+2Bh+3Ch^2+, \text{etc.};$$

next, with respect to x,

$$\frac{dY}{dx}=\frac{dy}{dx}+\frac{dA}{dx}h+\frac{dB}{dx}h^2+\frac{dC}{dx}h^3+, \text{etc.}$$

But the differential coefficients of like powers of h are identical, because the same thing has been differentiated in each case.

Hence, $A=\frac{dy}{dx}$, $B=\frac{dA}{2\,dx}$, $C=\frac{dB}{3\,dx}$;

that is, $A=\frac{dy}{dx}$, $B=\frac{d^2y}{dx^2}\cdot\frac{1}{2}$, $C=\frac{d^3y}{dx^3}\cdot\frac{1}{2.3}$, etc.

Hence we have Taylor's Theorem,

$$Y=y+\frac{dy}{dx}\cdot\frac{h}{1}+\frac{d^2y}{dx^2}\cdot\frac{h^2}{1.2}+\frac{d^3y}{dx^3}\cdot\frac{h^3}{1.2.3}+, \text{etc.}$$

If h be negative in the original $F(x+h)$, the signs of the terms having the odd powers of h will be negative, in *general* expression, but such terms may have a *particular* positive value.

1. Place $Fx=y$, namely, $15x+x^2=y$, into Taylor's Theorem, or as it will be when x becomes $x+h$.

Ans.

$$F(x+h)=15x+x^2+(15+2x)\frac{h}{1}+(2)\frac{h^2}{1.2}+0+0, \text{etc.}$$

2. Place this $Fx=y$ in Taylor's Theorem, viz.:

$$5x + 19x^2 - 3x^3 = y.$$

$$F(x+h) = (5x + 19x^2 - 3x^3) + (5 + 38x - 9x^2)\frac{h}{1} +$$
$$(38 - 18x)\frac{h^2}{1.2} + (-18)\frac{h^3}{1.2.3} + 0 + 0 +, \text{etc.},$$

where, on account of the factors h, etc., no term but the first has actual greatness when $h = 0$, but an initial or relative greatness. Now, since multiplying a function or a coefficient by a constant, does not affect the greatness of the variable in its relation to any laws or principles of change in the function or coefficient, or to its sign in the resultant of a term, we are able, for *most* practical purposes, to *dispense with the factors* $\frac{h}{1}, \frac{h^2}{1.2}$, etc., in the use of this theorem. And it is in this sense that we may have made an implied use of the theorem without a formal demonstration. The *coefficients* as factors of each term have, however, in themselves real amounts.

In noticing the example above, we might hesitate about the significance of $+ (-18)$, but it is of course $= -18$; hence an important notice that the plus of a *general notation* may be reversed by special considerations. In the $\frac{dy}{dx}$ and $\frac{d^2y}{dx^2}$ above, i. e., in their equivalent in the particular quantities, the actual sign of the resultant of a term cannot be known without hypothesis for the value of x, although the notation gives $+$ before the general term.

3. Place $(a - x)^3 = y$, in Taylor's Theorem, in its state which has just been passed by in consequence of the last growth of x. Making h the decrement:

$$F(x - h) =$$
$$(a - x)^3 - 3(a - x)^2 . \frac{h}{1} + 6(a - x) . \frac{h^2}{1.2} - 6 . \frac{h^3}{1.2.3}.$$

4. Do the same with $x^5 = y$.
5. Do the same with $-x^3 = y$.
6. Present, in Taylor's Theorem, the variable taking an increment, some terms of the function of x, viz.:

$$\frac{(ax-b)x}{b+x^3} = y.$$

7. Present, in Taylor's Theorem, a few terms of $\frac{a}{x^{\frac{1}{2}}} = y$.

8. The same required of $\frac{a-x^2}{x+1} = y$.

Taylor's Theorem has failing cases.

---

100. Says Professor Playfair:

"A single analytical formula in his (Brook Taylor's) Method of Increments has conferred a celebrity on its author which the most voluminous works have not often been able to bestow. It is known by the name of Taylor's Theorem, and expresses the value of a variable quantity in terms of the successive orders of increments. . . . If any one proposition can be said to comprehend in it a whole science it is this, for from it almost every truth and every method of the new analysis may be deduced. It is difficult to say whether Taylor's Theorem does the most credit to the genius of its author or the power of the language which is capable of concentrating such a vast body of knowledge in a single expression. This Theorem was first published in 1715."

SECTION XI.

THEORY OF MAXIMA AND MINIMA.

101. The most marked characteristic of a function, whether of one or more variables, is a maximum or minimum value of it; it is so marked as at once to arrest earnest attention; but yet, it is only an incidental or particular condition among its general values, or in its general nature.

102. A function is at a maximum value, or, in abridged language, is at a maximum, when it has such value as will be diminished in its nearest and earliest change, if the variable, having the value answering to that condition, be either increased or diminished the least amount; so that we must have

$$Fx > F(x+h),$$

and $$Fx > F(x-h).$$

103. A function is at a minimum when it has such value as will be increased in its earliest and nearest change, if the variable be either increased or diminished the least amount; so that we have

$$Fx < F(x+h),$$

and $$Fx < F(x-h).$$

Here we may observe that we have in

$$F(x-h),$$

$$Fx,$$

and $$F(x+h),$$

three values of a function indicated in an indeterminately close rank of succession, and that

$$x - h,$$

$$x,$$

and $$x + h,$$

are the same and corresponding values of the variable.

We are referring to but one function, in the use of the expressions $F(x - h)$, Fx, and $F(x + h)$, and since we speak of Fx as at a fixed value, we should rather say $F(a - h)$, Fa, $F(a + h)$, meaning thereby three values of a function of a single variable, such as belongs to it for $x = a - h$, then $x = a$, then $x = a + h$; but in this we are using a *varying* x. Now $F(x - h)$, Fx, $F(x + h)$, are abbreviated modes of asserting the same thing, with a now *fixed* or *unvarying* x.

With the above explanation, we use x, etc., instead of a, etc., for the better preservation of associations connected with Taylor's Theorem. By this theorem we have the general developments or expansions for $F(x - h)$ and $F(x + h)$; between which, however, we will arrange Fx as an intermediate condition, with the value y; they are,

1. $F(x - h) =$

$$y - \frac{dy}{dx} \cdot \frac{h}{1} + \frac{d^2 y}{dx^2} \cdot \frac{h^2}{1.2} - \frac{d^3 y}{dx^3} \cdot \frac{h^3}{1.2.3} +, \text{ etc.}$$

2. $Fx = y.$

3. $F(x + h) =$

$$y + \frac{dy}{dx} \cdot \frac{h}{1} + \frac{d^2 y}{dx^2} \cdot \frac{h^2}{1.2} + \frac{d^3 y}{dx^3} \cdot \frac{h^3}{1.2.3} +, \text{ etc.}$$

Now since, as in the case of the Binomial Theorem (**60**),

to which the above are similar in respect to being series with the powers of h ascending, h may be taken so small that the *first or any term* having h as factor, must be greater than the sum of all succeeding terms, such term may be reasoned upon as *in itself* determining whether y is indicated as increased or as diminished by that *same any term*, and all its sequel, without our paying any regard to terms succeeding such one.

Concerning the factors of the several successive dif. coefs., viz.: $h, \frac{h^2}{1.2}, \frac{h^3}{1.2.3}$, etc., we may remark, that since h itself is supposed to be indeterminately small, we need not, in reference to the purpose answered by h, h^2, etc., be solicitous about any differences between them, nor whether either have as a divisor 1, or 2, or 6, or 24, etc.

If, then, $F x > F(x - h)$ and $F x > F(x + h)$, as in the case of $F x = \text{max.}$, then neither $-\frac{dy}{dx} \cdot \frac{h}{1}$ nor $+\frac{dy}{dx} \cdot \frac{h}{1}$ can indicate increase, which they would not do if these dif. coefs. were $= 0$, for then their signs become of no force.

Now we may be able to find the value x, if it has one, which verifies the following equation

$$\frac{dy}{dx} = 0,$$

Our general formulas, then, when $\frac{dy}{dx} = 0$, after eliminating this dif. coef., become,

4. $F(x - h) = y + \frac{d^2y}{dx^2} \cdot \frac{h^2}{1.2} - \frac{d^3y}{dx^3} \cdot \frac{h^3}{1.2.3} +$, etc.

5. $F x = y$.

6. $F(x + h) = y + \frac{d^2y}{dx^2} \cdot \frac{h^2}{1.2} + \frac{d^3y}{dx^3} \cdot \frac{h^3}{1.2.3} +$, etc.

Now it is evident that if $F x =$ a max., $\frac{d^2 y}{d x^2}$ must be negative, because there ought not to be indicated or realized any increase of y, in equations 4 and 6, $F x$ or y being by supposition the greater. If we test this second dif. coef. by applying to it the value or values of x already found, and find that it is negative, we shall already have verified a maximum for $F x$.

This second dif. coef., or any other, may have a particular value, the reverse of its sign by general notation, for its value must depend on that of x, and consequently the liability must be incurred of its value passing through zero, the sub-terms of such dif. coef. having of themselves $+$ or $-$ signs.

If, nevertheless, we find this $\frac{d^2 y}{d x^2}$ to be positive, as it should be in case $F x$ is a minimum, when of course $F x < F (x - h)$ and $F x < F (x - h)$, then we have verified a minimum for $F x$.

But the case may occur when, after verifying its value, we must find

$$\frac{d^2 y}{d x^2} = 0;$$

in such case we find the following to be expressive of this condition:

7. $F (x - h) =$

$$y - \frac{d^3 y}{d x^3} \cdot \frac{h^3}{1.2.3} + \frac{d^4 y}{d x^4} \cdot \frac{h^4}{1.2.3.4} -, \text{ etc.}$$

8. $F x = y$.

9. $F (x + h) =$

$$y + \frac{d^3 y}{d x^3} \cdot \frac{h^3}{1.2.3} + \frac{d^4 y}{d x^4} \cdot \frac{h^4}{1.2.3.4} +, \text{ etc.}$$

Now it is evident that $F x$ cannot be a maximum or a minimum unless $-\frac{d^3 y}{d x^3}$ and $+\frac{d^3 y}{d x^3}$ have the same sign, i. e., concur in indicating decrease of y in case of a maximum, or increase in case of a minimum, and they cannot concur unless each $= 0$, and then no amount is indicated; so this third dif. coef. becomes eliminated, and we procced to $\frac{d^4 y}{d x^4}$ with the same remark with which we approached $\frac{d^2 y}{d x^2}$ and should approach all *even* dif. coefs.

104. *In order, then, to determine the value or values of x, which render any proposed function a maximum or minimum, we must deduce the value or values of x from putting $\frac{d y}{d x} = 0$, and substitute such value or values successively in the succeeding dif. coefs., until we arrive at one which does not vanish, i. e., become zero; if this one be the* 3*d*, 5*th*, 7*th*, 9*th*, *etc., the value we have found will not render the function either a maximum or minimum; but if the dif. coef. not vanishing, be the* 2*d*, 4*th*, 6*th*, *etc., it will; the function being at a maximum if the dif. coef. proves negative; at a minimum if it proves positive.*

105. Since the variable may be found to have more than one value deduced from $\frac{d y}{d x} = 0$, the function may have more than one maximum, or more than one minimum, or any number of each: in which case, if the function is *continuous*, they may alternate in consecutiveness. In such case a maximum may be found for one value of x, and a minimum for another value of it.

106. It is obvious, on reflection, that when a constant is common to every term of a function as factor, such factor may be disregarded, in determining the value of x, at which a maximum or minimum occurs.

The following considerations are obvious and important in abridging the method of finding maxima and minima: —

107. Radical signs and indexes, affecting collectively every term of a function, may be disregarded.

108. There is no maximum or minimum of the function when x is infinite, because it has no *succeeding* value, nor when y is infinite, because it cannot be diminished by a finite quantity.

In an early section of this treatise, the rationale of maxima and minima was the most inductively demonstrated, with reference to selected functions; but there remained the advantages of formal instead of virtual differentiation, and the use of its nomenclature, to be pointed out in this section, if we would aim at the most explicit rules.

109. A variable x, on which a function y depends, may have its own maxima and minima, which may be found without the necessity of determining what $F\,y = x$; by deducing the value of x from the reciprocal of the first dif. coef. of the function put $= 0$, or from the first dif. coef. put $= \infty$.

110. When a function is of a fractional form, and has its *variable only in its denominator*, it is evident that the maximum value of the function corresponds with a minimum value of the denominator considered as a certain other function, and vice versa.

111. If the function is fractional in form, and the denominator constant, it is evident that the reciprocal of that denominator is the constant *factor* to the numerator. Hence, such a product is greatest when its variable factor is greatest, or the numerator alone.

112. When a dif. coef. is in the form of a fraction, it is evident that it must have the value zero, when its numerator has that value, if at the same time its denominator does not also reduce to the value zero.

113. Since it is always necessary to know the sign of a denominator in determining the value of a fraction, we may remark that when the denominator of a dif. coef. is an undisturbed second or greater *even* power, that denominator is always positive.

PROBLEMS.

Place, in Taylor's Theorem, this function of a single variable x, namely, $10\,x - x^2 = y$, first with an increment, second with a decrement, $\frac{h}{1}$, $\frac{h^2}{1\,.\,2}$, etc., being understood as successive *factors* of each term after the first.

$$F\,(x + h) = (10\,x - x^2) + (10 - 2\,x) + (-\,2) + 0;$$

$$F\,(x - h) = (10\,x - x^2) - (10 - 2\,x) + (-\,2) - 0;$$

also, $$F\,x = (10\,x - x^2) \pm 0.$$

Now if $F\,x$ is at a maximum or minimum, or first say if it is at a maximum, then neither $+\,(10 - 2\,x)$ nor $-\,(10 - 2\,x)$ can indicate increase of it; which they would not do if $10 = 2\,x$, or what is the *general* statement (by the notation of the theorem) if $\frac{d\,y}{d\,x} = 0$.

Now the $\frac{d^2\,y}{d\,x^2}$ being $-\,2$ in both cases, it intimates decrease in both cases, when its *factor* is allowed the *same* or as real a suggestion of being something as is suggested for the presumed change of the function's value.

Examine, as above, $x^2 - 10\,x$ for indications whether it has a maximum or minimum.

But $\frac{d^2\,y}{d\,x^2}$ might contain the variable and vanish, i. e., equal 0.

In such cases $\frac{d^3 y}{d x^3}$ takes the place in the continued reasoning of the first, and $\frac{d^4 y}{d x^4}$ takes the place of the second, and so on.

114. It is required to determine at what value or values of x, and of y, the following (functions of x) $= y$ have maxima or minima, if they have such.

1. $y = 25x - x^2$. Ans. y a max. when $x =$
   And when $y =$
2. $y = ax + bx^3$. Ans.
3. $y = x^2 - x$. Ans. $y =$ min. when $x =$
   And when $y =$
4. $y = 3x^3 + b$. Ans.
5. $y = x^2$. Ans. $x =$ min. when $x = 0$.
   And when $y = 0$.
6. $y = 3x^3 - 54x^2 + 315x + 5000$.
   Ans. $y =$ max. when $x = 5$, y being $= 5600$.
   $=$ min. when $x = 7$, y being $= 5588$.
7. $y = \frac{a}{x^2 - x^4}$. Ans.
8. $y = x^2 \times (b - x)$. Ans. $y =$ max. when $x = \frac{2}{3} b$.
9. $y = \frac{bx}{a} \times 2\sqrt{ax - x^2}$.
   Ans. $y =$ max. when $x = \frac{3}{4} a$.
   And when $y =$
10. $y = 6x^{\frac{1}{2}} - x^2$. Ans. $y =$ max. when $x =$
    And when $y =$
11. $y = \frac{4x^3 + 2a}{5x^2}$. Ans. $y =$ min. when $x = a^{\frac{1}{3}}$.
    And when $y = \frac{6}{5} \times a^{\frac{1}{3}}$.
12. $y = 60 + x^3 - 3ax^2 + 3a^2x - a^3$.

In this instance, if $\frac{dy}{dx} = 0$, $x = a$, but when $x = a$, $\frac{d^2 y}{d x^2} = 0$, and $-\frac{d^3 y}{d x^3} = -6$, and $+\frac{d^3 y}{d x^3} = +6$; so that, since there is not a concurrence of the values of the third dif. coef. in sign, and this third dif. coef. cannot $= 0$ in accordance with the rule which is here sustained, there can be no maximum nor minimum.

13. $y = (x - a)^4$. Ans.

14. $y = (x - a - b - c)^{2n}$, n being a whole number.
Ans.

15. $y = (b - x)^{2n+1}$, n being a whole number.
Ans.

16. $y = (b - x)^5$. Ans.

17. $y = b + (x - a)^{\frac{2}{3}}$.

This is a case of an exception to the rule, for all the dif. coefs. become infinite when $x = a$, and y is then $= b$. There is a maximum when $x = a$, because $F x > F(x - h)$ and $F x > F(x + h)$, which may be verified by algebraic methods, because Taylor's Theorem fails. But if the exponent be greater than 1, and less than 2, and its denominator odd, $\frac{d y}{d x}$ will not be infinite in a function of which the root is $x - a$, at the value of $x = a$, but there will not be a maximum or minimum because $\pm \frac{d^2 y}{d x^2}$ becomes imaginary for its negative sign; as in

18. $$y = b + (x - a)^{\frac{1}{2}};$$

where y has no max. or min.

19. (a.) How great can $y = 8x - x^2$ be?
Ans.

(b.) How small can $y' = 80 + x^2 - 10x$ be?
Ans.

(c.) What is the value of x if we put $y = y'$?
Ans. Imaginary.

Can we then, indiscriminately, by hypothesis, make any two functions of the same variable equal to each other; that is, their difference $= 0$?

20. Has the variable x a minimum in $y = (x^2 - a^2)^{\frac{1}{2}}$? Ans.

21. $y = (a x - x^2)^{-1}$.

22. $y = 20 + (6 - x)^{\frac{1}{3}}$.

23. How great is x while positive, and y is increasing the fastest in

$$y = 30 x + 180 x^2 - 20 x^3?$$

Make $\frac{d y}{d x}$ a maximum, or $\frac{d^2 y}{d x^2} = 0$, and determine x.

24. $y = (x - a)^5$.

25. $y = (x - b)^6$.

26. $y = x^7 - 7 c x^6 + 21 c^2 x^5 - 35 c^3 x^4 + 35 c^4 x^3 - 21 c^5 x^2 + 7 c^6 x + 175 a b$.

27. $y = (x + a)^6$. Ans. A minimum when $x = -a$.

28. $y = (x + c)^7$.

The above functions are offered to bring into use and to verify more of the conditions of the demonstration with regard to dif. coefs. after the second, than is commonly required. They will show the utility of Taylor's Theorem, as the foundation of the demonstration, to be remarkable.

SECTION XII.

PROBLEMS FURNISHING EXPLICIT FUNCTIONS OF ONE VARIABLE; FOR DETERMINING THEIR MAXIMA AND MINIMA.

115. This section will present a collection of problems. It is not to be expected that every value of a function or variable, which may be algebraically determined, can have a rendering or practical use within the conditions of any problem; or that the conditions and elements of the problem can be restated for accommodation of all such algebraically determined results. In articles **6** and **7** we found that this is not possible; we shall have repeated occasions to verify the same impossibility.

Of the *four* conditions of value for a function in relation to those of the variable mentioned in article **97**, we shall find that generally not more than *one* is available for any significance within the conditions of the practical economy of such problems; but occasionally two are.

---

1. (*a.*) The present problem is a common algebraic one: *A* and *B* set out from two towns, distant 247 miles from each other, and travelled the direct road till they met. *A* went 9 miles a day, and the number of days at the end of which they met, increased by the number of miles *B* went per day, was 31. Required the number of miles *B* went a day. Ans. 23.36 miles, or — 1.36 miles.

Here the negative result must be rejected, for although it might be executed as miles, it evidently could not as days.

(*b.*) The same problem as for the calculus: *A* and *B* set out from two towns, distant 247 miles from each other, and travelled the direct road till they met. *A* went 9 miles a day, and the number of days, at the end of which they met, increased by the number of miles *B* went a day, was the least possible, on the conditions. Required the number of miles *B* went a day. Ans. $6\frac{71}{100}$.

(*c.*) Required the number of days at the end of which they met, increased by the number of miles *B* went a day, as by the last condition. Ans. 15.65.

2. A certain company at a tavern had a reckoning of 143 shillings to pay, but 4 of the number being so ungenerous as to slip away without paying, the remainder settled the bill after the landlord had thrown off 10 shillings from the amount, and it was found that the original company was such, or within a fraction of such, that a payer's portion was increased the greatest amount it could be for any number for the original company, greater than four. Required the number as diminished by any such fractional man.

3. (*a.*) A man, *F*, is one of a crew of a returned fleet of fishing vessels, which are the same in number as the men of the like crews of each vessel; and he receives his equal share of 960 dollars of bounty money. His own crew with himself expend 24 dollars for dining together. It is required to adjust the number of men to a crew, for the condition that, by this receipt and disbursement, *F* finds the least money left as his own.

Ans. 80 men; *F* pays 15 cents more than his share of the bounty.

(*b.*) It is required to adjust the number of men for a crew for the condition that the bounty money should just pay *F* and his companions, if he had any, for their dinner; also required the price of the dinner to *F* for this condition.

(*c.*) It is required to describe the fleet in number of men and vessels, if it may be called a fleet, for the condition that F, after paying for the dinner, comes off with 936 dollars.

4. A certain pin factory has on hand ready for sale, except in the packing, 3,200,000 rows of pins as they are commonly stuck in papers. From these pins 100 boxes are to be made up, if possible, for sending away, each box containing packages, each package containing papers, and each paper rows, each row pins, the same number. Now the number of individual pins not included in this lot was the greatest possible. Required the entire number made; the number to be sent away, the number that will remain, and the number in a row.

Ans. In part, 64,000,000; 20 in a row.

5. Required the numerical quantity which exceeds its second power the most. Ans. $\frac{1}{2}$.

6. Required the numerical quantity of which twice the second power exceeds thrice its third power the most.

Ans. $\frac{4}{9}$.

7. Two vessels, A and B, were freighted each with 500 or a tons of coal. On the passage, the vessel A having sprung a leak, a certain number of tons were transferred from A to B; both cargoes were sold for one 35th (one bth) as many dollars per ton as there were tons so transferred. Now the mutual product of the number of dollars of the proceeds of the two unequal cargoes was the greatest possible. Required the number of tons transferred. Ans. 353 + tons.

The function of the variable which is put at the maximum is

$$\left((a - x)\,\frac{x}{b}\right) \times \left((a + x)\,\frac{x}{b}\right),$$

and it will be seen that b exercises no power in determining x.

8. A speculator expended 900 dollars in the purchase of animals at an equal price each. After including with this lot 15 other animals purchased at the same cost each, at a subsequent time, he sold all for 1150 dollars, at equal rates each, when he found that he had gained the greatest possible profit on each animal. Required how many at first were purchased, and the profit on each.

Let $x =$ number purchased,

and $y =$ dollars profit on each.

$$\therefore \frac{1150}{x+15} - \frac{900}{x} = y,$$

$$\therefore \frac{250\,x - 13500}{x^2 + 15\,x} = y,$$

$$\therefore \frac{x - 54}{x^2 + 15\,x} = \frac{y}{250} = y'.$$

Ans. y and y' are at a maximum at the same value of x, viz., $x = 115.4$, the number of animals required.

9. Required the area and each side of the greatest right angled triangle which has the sum of its hypothenuse and base 18 inches.

10. Required how a rectangle must be restricted when it contains the greatest area for a constant sum of the four sides. Ans. It must be a square.

11. Required the sides severally of the largest right angled triangle of which the perimeter is 22 in any units of length.

12. (*a.*) A certain reservoir, containing an unspecified quantity of water, is receiving 153 casks of water per day; $9\frac{1}{4}$ times that cask full is distributed per day, to each of as many families as the cask holds gallons. This state of receipt and distribution is continued 45 times as many days as the measure holds gallons. How many gallons did

the cask contain, in case it was known that the reservoir gained and retained the most water possible.

Ans. $11\frac{1}{37}$ galls.

(*b*.) If the cask is supposed to be of some size, and then to grow larger, in the play of many suppositions, how great is it when, increasing at a fixed rate, it implicates the greatest gain of the reservoir's water. Ans. $5\frac{39}{74}$ galls.

(*c*.) Of what influence is the 45 in problem (*a*.) in the matter of influencing the above results? Ans. None.

13. (*a*.) A farmer has a triangular plain situated between three crossing public highways, so barren and pointed at one angle that it is not worth his while to make a fence enclosing all that point. The sides are 81, 74, and 15 rods. He concludes to run a straight fence on the side 15 rods long, and on portions of the other sides nearest this short one, and then run a fence across the lot parallel to the side 15 rods long. It is required to determine the length of this cross fence and the area of the fenced lot and of the unfenced lot, if all the fence, in number of rods long, bears the least possible ratio to the number of square rods fenced.

(*b*.) Releasing now the condition that the straight cross fence must be parallel to the side 15 rods long; required its length when the above ratio may be still smaller, and the smallest possible; then how many rods of the 81 and of the 74 are respectively fenced.

(*c*.) Required to determine if some portion of every possible plane triangle may not be cut off by a straight line and at each of its angles, and the law of all such lines, the perimeters being thus reduced to a minimum ratio in units of any name to the square units of area, for the six-sided figure thus produced, i. e., preserving some portions of the original sides in position.

Ans. The lines must all be tangent to the greatest inscribed circle.

(*d.*) Required to determine the nature of an original triangle when the hexagon thus produced may be the largest possible part of it. Ans. An equilateral triangle.

14. (*a.*) A farmer has a lot of land in the shape of a right angled plane triangle, of which the hypothenuse is a, the other sides b and c. Required to determine the length and breadth of the largest rectangular lot which can be laid out in it.

(*b.*) The sides of a plane triangle, not necessarily right angled, are respectively given, it is required to inscribe three largest parallelograms, the half of each as determined by a diagonal, being one identical portion of the original triangle.

(*c.*) And it is required to find the largest rectangle which can be inscribed in the above triangle, and having the equivalent of its own half in common with the halves of them.

15. Divide a into two such parts that, one part multiplied by the second power of the other, shall be a minimum.

16. If coin is conditioned to be cylindric in shape, which it always seems to be, how must it be proportioned that the greatest bulk may be united with the least surface, and consequently the liability to wear away by friction the least for a cylinder.

Ans. Diameter of face and thickness equal.

17. A turner, having the trunk of a locust tree in the shape of the frustum of a right cone, such that if the portion having the apex were restored, it would complete a cone 28 feet long in the axis and 16 inches in diameter at the base, is told to turn out the largest cylindric gate-post possible, whatever the length of the frustum might be. Required to determine what length of it he would use, and the diameter of the post required.

18. (*a.*) A turner is given a lignum-vitæ ninepin-alley spherical ball 4 inches in diameter, and told to turn out the largest possible right cone. Required its dimensions; and

the weight of the ball being uniform, required to determine the weight of his chips or waste, as compared with that of the cone. Ans. Height of cone $2\frac{2}{3}$ inches.

(*b.*) It is required to determine whether a section of such a ball, passing through the axis of such cone, shows a maximum triangle in a circle, — a problem which may be determined.

19. (*a.*) A milkman, *M*, delivered daily in New York some quarts of milk to each of some customers, on each of some streets, the quarts, customers, and streets being the same in number; and in addition he delivered 85 quarts in Brooklyn. Now *N*, another milkman, hearing of this, said he believed that himself furnished daily each of his own customers 8 more quarts than *M* did his, which *M* admitted; that, although they were all on Chatham Street, both agree that *N* had 3 times as many customers as *M* had on any street; *N* was, therefore, ready to wager that his own deliveries of milk, in the aggregate, were the greatest. The discussion grew warm, and they put the statements in writing as above. Admitting the claims of the latter, or giving *N* every advantage in the interpretation of numbers, does he win? Giving *M* every advantage, how much may his daily aggregate exceed *N*'s?

(*b.*) The next day *M* acknowledged that his delivery in Brooklyn amounted to only 72 quarts. How does the original bet stand on this hypothesis?

(*c.*) What is the difference of the aggregate deliveries of the two men (on the hypothesis of the 85 quarts delivered in Brooklyn) when they are the nearest alike? and who owns the difference? And, on this hypothesis, how many quarts do each deliver in all? and how many quarts to a customer?

20. (*a.*) A hound starts to catch a hare which is 22,000 feet, in their common line of motion, in advance of him; from which point the hare makes off 157 times as many

leaps as there were feet in each leap; during the same time the hound makes after the hare 2714 leaps, each of the same length as the hare's. Now, such was the length of the leaps that, when they were completed, the hound was brought nearer to the hare than had the leap been of any other length whatever; what was its length?

Ans. $8\frac{101}{157}$ feet.

(*b.*) Could the hound have caught the hare by any length of their common leap? Or could the function of the leap, which expresses the final distance of the animals, $=0$?

(*c.*) How far apart were they on completing the leaps $8\frac{101}{157}$ feet long? How far had each run?

(*d.*) Had both started when the hare was 2200 feet in advance, might there have been two distinct lengths of the leap, adopting either of which the hare would have been caught, and in what two places? And what length of leap would have put the hound most in advance of the hare?

(*e.*) A hound starts to catch a hare, which is 22,000 feet in advance of him, and the hound makes 159 times as many leaps as there are feet in each; while the hare, starting at the same instant, makes 2814 leaps, each of the same length as the hound's. Adjust the leap to the greatest success or gain of the hare. Distance apart then, and the travel of each, what?

(*f.*) Supposing the animals move uniformly, and during the same time, according to a rate that is to produce a result of distance specified in problem; when the hare has leaped 100 rods, where is the hound?

21. (*a.*) An incendiary, escaping arrest, travelled by railroad cars $3\frac{8}{9}$ hours at a certain rate, when an accident happening, tending to delay the cars indefinitely, he resorts to horses upon a highway by the side of, and continuous with, the railroad. He thus rides at one third the rate at which he had gone by cars, and as many hours as leagues

per hour, and successively on as many horses as hours on each, when he judges it best to conceal himself. The next day an officer travels in pursuit from the same point of starting, and proceeds 6 times as far as any one horse had carried the fugitive, and $7\frac{1}{3}$ leagues more, and the pursuit ends; now, whether he arrives at the locality of the fugitive or not, may depend upon the rate at which the latter travelled, say by cars, per hour. Required the rate or rates when the officer accomplished as much distance as the fugitive.

(*b*.) In case that by any rate of travel of the fugitive by cars, we find the pursuer accomplished as much distance as the fugitive, it is required to determine which, by any possibility, may have accomplished the more distance, and how much more.

(*c*.) Required the limits of rate of travel of the fugitive by cars, by which he accomplishes the more distance; also the limits by which the pursuer accomplishes the more.

Ans. At three different values of x in $\frac{x^3}{27} - \frac{6x^2}{9} + \frac{35x}{9} - 7\frac{1}{3} = y$, is $y = 0$; at $x = 5$, y is at a max.; at $x = 7$, y is at a min.

22. The square root of a certain numerical quantity is taken from 57, and the remainder multiplied by that square root, and the product is a maximum. Required the quantity. Ans. $812\frac{1}{4}$.

23. A and B own 1500 square rods of land, and also 5 equal square lots lying together, and they propose to divide all their land between themselves. In consideration of the quality of the land, A agrees to accept and B to grant a lot, a side of one of the lots in one dimension and 100 rods in the other. Whence B's share was found to be the smallest tract he could possibly have, for any size of those lots whatever. Required the dimensions of B's share, and of one of the lots. Ans. B's share 1000 square rods.

24. There are two level lanes, which are straightly and perpendicularly walled at each of their sides; their widths are respectively 13 and 17 feet. They meet at right angles. A straight pole, of which the diameter may be called nothing, is to be carried level past this corner on the shoulder of a boy. Required the length of it when it is the longest possible. Ans.

Here the minimum length of pole is the *logical* maximum, in the economy of the problem.

25. A right cone is one of which the axis is perpendicular to the base. The base of a right cone is 8 inches, its height is 14 inches. Required the diameter of the largest sphere which it can enclose.

26. (*a.*) How far apart must a person, whose feet are b or 10 inches long, place the foremost end of his feet, while his heels are together, that the area of the base, on which he may be said to stand, may be the largest, and, therefore, the most secure as a general support?

(*b.*) But since his heels cannot in strictness be placed on one point, it is required to determine the above question, with the allowance that his heels may be a or 15 inches apart, but we will now condition that the feet be symmetrically situated, with reference to the line which joins the heels.

The solution of this problem will embrace the previous one if we put $a = 0$.

Let $\qquad x =$ the distance required

then, $$x = \frac{a}{2} \pm \sqrt{2b^2 + \frac{a^2}{4}}.$$

The ambiguous sign of the above result indicates that the principle involved does not discriminate that the weight of the body bears on the heels more than on the toes, and so tolerates the condition that the foremost ends of the feet may be nearest together, as well as the heels, which is true.

27. A company of 90 men was formed, in 1849, for mining in California, on equal shares, but before actually commencing labors they are induced to admit more members into the partnership. After working one day the whole company take 9 pounds of gold dust, and the second day take as many pounds as those *new* members number. On the third day, "prospecting," the whole party take no gold, but lose, *each man*, by thieves $\frac{1}{144}$th as many pounds of gold as those persons numbered who last joined them, when the company conclude to settle up, to allow the members to labor individually. It is required to adjust the number of those *latest* members to the greatest luck or good fortune of an individual of the whole for the three days, and to determine how much gold was a share.

Let $90 = b$, $9 = a$, and $\frac{1}{144} = c$,

and $x =$ the new members;

then one member's share is

$$\frac{a+x}{b+x} - c\,x = y,$$

$$\frac{d\,y}{d\,x} = \frac{b-a}{(b+x)^2} - c = 0 \text{ in case of a max. or min.,}$$

$$\therefore x = \pm\sqrt{\frac{b-a}{c}} - b = 18 \text{ or } -198,$$

$$\frac{d^2 y}{d\,x^2} = \frac{-b^2 + a\,b - b\,x + a\,x}{\frac{1}{2}(b+x)^4},$$

which is negative under the values which these constants are known to have; indicating a maximum for y while x is positive at 18, and a minimum for y when x is made -198.

However, x with the value -198, has no application to the language of the problem *as enunciated.*

116. This occasion is taken to remark that differential coefficients are not always to be necessarily regarded as mere numerical amounts or ratios. They are functions as well, and it is useful to read them in the language of a problem's special kinds of quantity. Thus, in regard to the problem of the 90 California miners as first stated, the first dif. coef. of such function as always expresses one actual laborer's share of gold dust, for every possible number of new members, which coef. is

$$\frac{dy}{dx} = \frac{b-a}{(b+x)^2} - c,$$

when read as a *derived function* with the significance of the quantities preserved, may be as it follows after this preamble:

One actual laborer's share in pounds will always be found to vary (as depending on the variation of the number of new men) just as the following supposed share, compared with 1 pound, will vary, viz.:

From as many pounds of gold dust as the original company numbered men (90), take what the workers obtained the second day (9), and divide the remainder among the actual workers as their number would be after each worker had withdrawn, and put in his own place a company equal to the whole workers, and then take away from each such share $\frac{1}{144}$th part of a pound.

28. (*a.*) A and B set out at the same time, from places 320 miles apart, and travel to meet. Each travels uniformly at his own rate, and the number of hours at the end of which they meet, is equal to one half the number of miles which B goes per hour. May there be any number of miles which B may go per hour, according to which any possible

difference between their rates per hour may be a maximum or minimum?

(*b.*) Less than at what rate per hour can they not go, when they both travel alike? If they travel at the same rate, what is that? How much slower may A go than B? B than A? Ans. Infinitely.

(*c.*) What is the rate of each, if they differ 3 miles per hour?

Ans. Ambiguous, because it is not hypothecated which goes the faster.

(*d.*) Required the rate of each when A goes 3 miles per hour faster than B, and the rate when B goes 3 miles faster than A.

29. Two straight lines, $A\ B$ and $A\ C$, of indefinite length, meet at the point A at right angles. It is required to determine the length of the shortest hypothenuse that shall pass through a given point situated in the plane of those two lines, at the perpendicular distance a from $A\ B$, and b from $A\ C$, and complete a right angled triangle.

Let $x =$ that part of $A\ C$ not equivalent to a;

then $\frac{a\,b}{x} =$ that part of $A\ B$ not equivalent to b, because $x : b : a : \frac{a\,b}{x}$;

let $y =$ the hypothenuse,

then $y = \sqrt{(a + x)^2 \pm \left(b + \frac{a\,b}{x}\right)^2}$,

$$y' = (a + x)^2 + \left(b + \frac{a\,b}{x}\right)^2,$$

$$\frac{d\,y'}{d\,x} = 2\,a + 2\,x - \frac{2\,a\,b^2}{x^2} - \frac{2\,a^2\,b^2}{x^3};$$

y is a minimum when y' is.

This problem is more general in case the angle at A is any, and the hypothenuse is called the unspecified side, but requires trigonometry.

30. Required the area of the greatest right-angled triangle which has 11 inches for its hypothenuse; also, which has a for its hypothenuse.

31. There is a cylindric tin pail without a cover, of which the bottom is 9 inches in diameter, and height 8 inches. Required to know if another pail can be máde that may hold as much water with less sheet tin, and how much less; and required the rule of proportion between these dimensions, both when having a flat cover and when without one.

32. What decimal fraction exceeds its cube more than any other numerical *quantity* whatever exceeds *its* cube?

Ans. .577 +.

33. Divide 25 into two such parts, that the product of the second power of one part by the third power of the other, may be larger than any other product of its parts at those powers.

Ans. 15 and 10, the larger to be of the third power.

34. (*a.*) Two farmers, A and B, laid out for themselves each a farm of equal territory and rectangular shape; a straight line drawn from a corner of A's farm across it, and meeting a side 68 rods from that corner which is diagonally opposite the corner of starting, is 152 rods long. The remainder of the side, a part of which is the 68 rods, is of the same length as one side of B's farm. Required the length and breadth of each farm when they are the largest they can be upon these conditions.

(*b.*) If B's farm, by any dimensions we may adopt for it, is square, required the length of one side; and the two sides of A's.

35. A farmer, having at first 80 dollars, sold 3 times as many bushels of potatoes, as he sold them at in cents per

bushel, and then purchased 246 bushels of corn, each bushel at the price of a bushel of potatoes just sold, and has left a least possible or a greatest possible amount of money. Required whether greatest or least, and the prices of the potatoes and corn per bushel.

36. From the equator a ship sailed north 50 times as many hours as she sailed miles per hour; thence she sailed south 1000 hours at the same rate, and the result was the least possible gain to the north. Required how many miles per hour she sailed, and how far she is from the equator.

37. (*a*.) Supposing the Boston and Worcester Railroad and the Old Colony Railroad to run from Boston at right angles to each other, and the roads straight, and that a train of cars on the Worcester road, 19 miles from Boston, is ready to start, headed for Boston, at 32 miles an hour, and a train on the Old Colony is ready to leave Boston at the same instant, at 20 miles an hour, how far from Boston will each train be when they are nearest together by a straight line across the country, and how long after starting, and how far apart then?

Ans. In part, the train on the Worcester road $5\frac{125}{352}$ miles from Boston.

(*b*.) Repeat the problem, with the conditions all the same, except that the Old Colony train is to have 25 miles the start of the former condition, and see if any indication is offered that the occurrence we are watching for must have already happened, or would occur in the future, on each train reversing its direction, and where will the condition exist, and how long after starting from this position.

38. It is required to find a numerical quantity such that if from 9 times itself its second power be subtracted, the remainder will be equal to 3 times the quantity plus another sum; what is the quantity when this other sum is the *greatest* or *least* possible, and which of the *two?*

39. The sum of two quantities is 22; the second power of one added to twice the second power of the other, is a maximum or minimum; which? Required the numbers.

40. The difference between two quantities is 10, and the difference of their third powers is a maximum or minimum. What are they?

41. A wholesale druggist bought 542 (a) pounds of a drug at \$3.57 (b) per pound, and sold from it, a part at the same number of cents per pound as equals the number of pounds not sold, and the whole amount of the profit or loss on this sale was the greatest for any quantity sold; at what price was that portion sold per pound, and what the amount of the profit or loss on that portion as a whole?

Let $x =$ cents per pound of that sold;

$\therefore x =$ the number of pounds not sold;

$\therefore a - x =$ the pounds sold;

$\therefore \pm x \mp b =$ profit or loss per pound on that sold.

Let $y = (a - x)(\pm x \mp b) =$ all the profit or loss;

$\because y$, i. e., $\pm (a + b) x \mp x^2 \mp a b =$ max. or min.,

$$\therefore x = \frac{a + b}{2} = \$4.49\tfrac{1}{2},$$

$$\frac{d^2 y}{d x^2} = \pm 2;$$

$\therefore y$ is a maximum for profit, and minimum for loss, at the same value of x.

Since $x > b$, the sale is known to be at a profit, $\therefore y =$ max., because the negative sign of second dif. coef. becomes adopted.

NOTE. In the use of the doubtful sign $\pm$, care should be taken to preserve throughout the above work their proper correlation; the upper all belonging together in the logical relation, or the under.

42. In a certain country may be found a factory which turns out toys. Factory No. 1 makes a toy, which it packs in a paper; and with regard to this factory, one such package may be indifferently called a *paper*, *package*, *case*, or *box*. Factory No. 2 puts 2 of its toys in a paper, 2 papers in a package, 2 packages in a case, 2 cases in a box. Factory No. 3 does the same, except it puts 3 toys in a paper, 3 papers in a package, 3 packages in a case, 3 cases in a box. Factories No. 4, 5, 6, etc., put 4, 5, 6, etc., respectively, toys up by the same rule of packing; — each factory commencing with one more toy, paper, package, and case than its predecessor, and so on with factories indefinitely. Now, in one of the two packing rooms in each factory, the last stage is the packed case, in the other, the last stage is the packed box. Suppose that in the case room of each factory, there are 144 packed cases; enough of these are carried into the box room to make 3 boxes, if possible. These boxes are then collected and shipped. Remaining in the case room of which factory, is there the greatest number of individual toys? In which remain none? and between what factories is the difference the greatest, in the number of toys left in the case rooms?

Ans. In the 36th is the greatest number; in the 48th none; the difference the greatest between the 23d and 24th; in the 23d, toys 912,525; in the 24th, 995,328; in the 25th, 1,077,375; difference between 23d and 24th, 82,803; 24th and 25th, 82,047.

It is obvious that the problem cannot have practicability under such generalizations of the constants, as would render the above results fractional. Thus, if 144 were to be replaced with 141, the 36th in the above answer becomes $35\frac{1}{4}$.

43. The sides for constructing a quadrilateral plane figure, are consecutively 27, 19, 42, and 31 units of length,

any three of which are greater than the fourth; it is required to determine the diagonal meeting the sides 19 and 42, and that meeting the sides 42 and 31, when the area of the quadrilateral is the greatest possible. Required that area in square units of the same name.

44. A man having a sum of money, gave 90 dollars of it to a charity fund, and distributed the remainder of it to the same number of deserving orphans, as dollars to each, one of whom was T. Now, there were some schools close by, as many in number as T received dollars, each school having as many boys, and each boy *owning* as many dollars as T received, or *owing* as many, for we did not hear distinctly, but consider that it might have been either way. Required to determine the relation between the sum of dollars originally possessed by the man, and the possessions in dollars, or debts of these boys, and their relative rates of variation for all possible sums.

Let $x =$ the man's sum;

and $y =$ all the boys' sum;

$$\therefore y = \pm (x - 90)^{\frac{3}{2}};$$

$$\therefore \frac{dy}{dx} = \pm \frac{3}{2} (x - 90)^{\frac{1}{2}}.$$

Here $\frac{dy}{dx}$ may $= 0$, but y has no maximum or minimum, because $d F (x - h)$ is imaginary when $x = 90$.

45. A man having a number of dollars in a purse, put 10 dollars of it into a Savings Bank, and having divided the remainder into as many parts as he put dollars in a part, proceeded, having other money in a pocket book, to purchase some articles, which were just 22 in number less than the original number of dollars in the purse, paying for each article a sum equal to one of those parts. Required

to trace all the relations between the number of dollars in that purse, and the value of all those articles purchased.

The function $(x - 22)\sqrt{x - 10} = y$, has the value zero when $x = 22$, and when $x = 10$, and an algebraic maximum and minimum when $x = 16$, which value is incompatible with the conditions. After x exceeds 22, y is practicable to infinity in its positive values.

Definitions.

117. A *pyramid* is a solid figure contained by planes, that are constituted between one plane called the base, and a point above it called the apex, in which they meet.

A *prism* is a solid figure contained by plane figures, of which two that are opposite are equal, similar, and parallel to one another; and the others are parallelograms.

A *parallelopiped* is a solid figure contained by six quadrilateral figures, whereof every opposite two are parallel.

A *right pyramid* has its apex perpendicularly over the centre of its base, when that base has such regularity as to have a centre, equally distant from the termination of each side of that base.

A *right prism* has rectangles for such of its sides as must be parallelograms.

A *right parallelopiped* is contained by no other plane figures but rectangles.

---

46. (*a.*) The base of a right pyramid is triangular, of which the sides are a, b, and c, and height is H; it is required to find the contents of the largest right prism

which can be contained in it, and each linear outline dimension of the same.

Ans. Contents, $\frac{1}{3} H \times \frac{4}{9} \times$ area of base.

Linear outline, $\frac{2a}{3}$, $\frac{2b}{3}$, $\frac{2c}{3}$, and $\frac{1}{3} H$.

(*b*.) The base of a pyramid is triangular, of which the sides are a, b, and c, and the perpendicular height is H; it is required to find the sides of the triangles that are parallel, containing the largest prism that can be contained in the pyramid.

(*c*.) The base of a right pyramid is rectangular a by b, and perpendicular height is H; required each linear outline dimension of the largest parallelopiped that can be contained in it.

(*d*.) The base of a pyramid is quadrilateral, the sides being a, b, c, and f; the area of the base A, and perpendicular height is H. Of such two like plane figures as are parallel to each other, and are detprminate, and bound the largest prism that can be contained in the pyramid, required the sides. Required also the ratio of the contents of the prism to those of the pyramid. Ans. Ratio $\frac{4}{9}$ths.

47. A hound discovers a deer 1800 (a) feet ahead of him: they both start the same instant, the deer in a direct line, pursued by the hound; the hound makes a leap each second, and 1 foot longer than the deer's, and makes 3 more leaps than the deer does in the time that the deer makes one less leaps than the number of feet in its length.

After taking 120 (b) leaps, the deer becomes arrested by the breaking of the crust of snow; at this instant the hound seeing the deer arrested, adds 2 feet to the length of his previous leap, and makes 7 in the time that he made 6 before, which he continues till he comes up with the deer. Now, if the deer was occupied in extricating himself, or in resting during the greatest or least time possible, how long had been the deer's leap, to accommodate this condition?

Let $x =$ feet leaped by the deer at a leap,

and $y =$ number of leaps of hound after deer stops;

$\therefore b\,x =$ all the feet leaped by the deer;

$\therefore x + 1 =$ feet of hound's leap at first;

$\therefore \frac{x+2}{x-1} \times b\,(x+1) =$ h.'s dis. attained when d. stops;

$\therefore a + b\,x =$ whole distance for hound to go;

$a + b\,x - \frac{x+2}{x-1}(b\,x + b) =$ to be leaped by h. after d. stops;

let $\frac{7}{6} = c$;

let $e = a - 4\,b$;

and $g = + a + 2\,b$;

$$\therefore y = \frac{a + b\,x - (b\,x + b)\,\frac{x+2}{x-1}}{c\,(x+3)};$$

$$y = \frac{e\,x - g}{x^2 + 2\,x - 3} \times \frac{1}{c};$$

let $y' = \frac{e\,x - g}{x^2 + 2\,x - 3}$;

$\therefore \frac{d\,y'}{d\,x} = \frac{2\,g\,x - e\,x^2 - 3\,e + 2\,g}{(x^2 + 2\,x - 3)^2} = 0$, when $y =$ max. or min.;

$\therefore$ the numerator $= 0$;

$\therefore x = \frac{g}{e} \pm \sqrt{2\,g - 3\,e + \frac{g^2}{e^2}} = 12.66$, or $- 9.46$;

$$\frac{d^2 y'}{d x^2} = \frac{(x^2 + 2x - 3)(2g - 2ex) - (2x + 2)(2gx - ex^2 - 3e + 2g)}{(x^2 + 2x - 3)^4};$$

now $$y = y' \times \frac{6}{7}.$$

Now, y is a max. when y' is, as is evident, and $\frac{d^2 y'}{d x^2}$ is negative when $x = 12.66$; hence a maximum which is an answer, and the negative value of x, is not within the significance of the language of the problem.

SECTION XIII.

COMPLETE HISTORY OF FUNCTIONS.

118. It is useful to become acquainted with the methods of fully examining the entire history of a function of one or more variables, in respect to the range of values which the function and its variable may sustain, and to their mutual dependence. Attention should also be paid to every constant in its influence on the function. All the prominent characteristics of value, and the rates of change of value, should be noted, by special regard to the value of dif. coefs.

A function should be tested for the values $+\infty$ and $-\infty$, and for how many times or successions it has either, and at what values of the variable respectively; whether it passes from $+\infty$ to $-\infty$, or from $-\infty$ to $+\infty$ with instantaneity, so to speak, as when we should examine $y = x^{-1}$ and $y = x^{-2}$, and observe the marked difference in respect to $x > 0$ or $x < 0$, i. e., while x passes the value 0.

A function should be tested for what it becomes when

the variable $= 0$, and for whether it has more than one value for the variable $= 0$; it should be tested for the variable $= +\infty$ and $= -\infty$.

A function should be tested for its value, zero, for how many times it has such, if at all; for at what value or values of the variable it has such, whether it passes *through* such value or not, and if through, at what rate of change.

A function should be tested for maxima and minima, for how many times it has either, and at what values of the variable; the same dif. coefs. may be employed to determine with facility, whether the *variable of itself* has maxima or minima; and if so, at what values of the function.

Since the constants that occur in a function are quantities, with which both the variable and the function are most concerned, or with which they are most compared, and in consequence are liable to exhibit marked characteristics when they become equal to or pass the value of such constants, tests should be applied for the values of both the variable and the function, when either is equal to such constants, but sometimes to a combination of such constants, as a product, sum, or quotient.

Functions of a single variable should be studied with reference to interrupted values, i. e., to losses of continuity.

A function should be tested for the maximum or minimum of its rates of change of value.

A function should be tested for proof of symmetricalness, or having the same value when its variable is a given amount less and more than some specific quantity or 0; and the variable be tested for having the same value when the function is a certain amount less and more than a specific quantity or 0.

It will be necessary to be prepared for cases in which neither the function nor variable can have any real values whatsoever.

The nature of these inquiries may be shown by an instance:

1. Let it be required to determine the range of all possible parts into which the number 50 might be divided, both when one of the parts is not greater than 50, and when it is any number whatsoever, estimated by algebraical equivalents; or,

Required the chief points in the history of the function,

$$y \text{ or } Fx = 50x - x^2 = (50 - x)x.$$

Fx will be found to have a maximum when, and only when, $x = 25$, Fx being 625, for there is but one value of x at which there is any maximum. It can have no minimum, because $\frac{d^2y}{dx^2}$ is always -2, and of course such when $x = 25$.

Fx cannot be so great as 625 at any other value of x than 25, because in the equation,

$$50x - x^2 = 625,$$

x will be found to have but one value, viz., $x = 25$; Fx, therefore, cannot $= +\infty$ in value.

Fx may $= -\infty$ both when $x = -\infty$, and when $x = +\infty$, for, solve the equation,

$$50x - x^2 = -\infty,$$

$$\therefore x = 25 \pm \sqrt{\infty + 625} = \pm\infty.$$

Fx has the value 0, both when $x = 0$ and when $x = 50$, for, solve the equation,

$$50x - x^2 = 0,$$

$$\therefore x = 0 \text{ or } 50.$$

Such are the general outlines of the history of possible

values of $F x$ and of x. Commencing now, with x algebraically the smallest conceivable, or $= - \infty$, $F x$ being also $- \infty$, we have,

$$y = 50 x - x^2 = - \infty,$$

and
$$x = - \infty,$$

and
$$\frac{d y}{d x} = + \infty;$$

because $\frac{d y}{d x} = 50 - 2 x = 50 - (2 \times - \infty) = 50 + \infty = \infty$.

Since the first dif. coef. $= \infty$ when $x = - \infty$, $F x$ is indicated about to increase or become a smaller negative, at an infinite positive rate. And since $\frac{d^2 y}{d x^2}$ is always $- 2$, such rate of increase is ever to diminish, and we know that $50 - 2 x$ grows less as x increases. These conditions continue till $F x = 0$ and $x = 0$, when $\frac{d y}{d x} = 50$; hence $F x$ passes through zero, increasing 50 times as fast as x, and beginning now to have positive values, goes on till $F x = 625$, and $x = 25$, when $\frac{d y}{d x} = 0$. $F x$ now returns to have a less value, till it $= 0$, x then being 50; in diminishing, $F x$ passes through the value 0, while

$$\frac{d y}{d x} = 50 - 2 x = 50 - 2 \times 50 = - 50;$$

which indicates that $F x$ passes again through zero, diminishing 50 times as fast as x: so $F x$ goes on to $= - \infty$, already shown, when $x = + \infty$ and $\frac{d y}{d x} = - \infty$, as shown.

Again, $F x$ is symmetrical in its history before and after its maximum; i. e., while x is any amount greater or less than 25, $F x$ has the same value. This may be verified thus:

Let $$x = v + 25,$$

and $$x = 25 - w;$$

$$\therefore 50(v+25) - (v+25)^2 = 50(25-w) - (25-w)^2,$$

whenever $w = v$, because, on such condition, the members of the equation reduce to identity.

If the given function should be as general as

$$a x^n - b x^{2n},$$

this characteristic of symmetricalness could be shown.

2. Let it be required to determine whether the equation of the Second Degree between two variables, x and y, viz.:

$$50 x - x^2 - y = 0,$$

involves the *elliptic*, *hyperbolic*, or *parabolic* condition. (Arts. **50**, **51**, **52**.) Ans. The parabolic.

3. It is required to determine whether $F x = -x^3 + 3x^2 + 24x - 85$, has a maximum and a minimum at any values of x, and to explain how it can be $+\infty$ and $-\infty$ also.

4. It is required from $F y = x$, viz., $50y - y^2 = x$, to obtain some function of x equal to y, and recount its history.

5. Required the complete history of $y = \frac{1}{x}$.

6. Required the complete history of $y = \frac{8 - x}{15 + x}$.

7. Required the complete history of $y = \frac{a + x}{b - x^2}$.

SECTION XIV.

PRINCIPLES AND PROBLEMS RELATING TO PROJECTED BODIES.

119. The height of *any point* in the course which a dense body like a stone or a mass of water takes, when thrown in any direction whatever, near the earth and through a medium no denser than the air, and at velocities not exceeding 300 or 400 feet per second, may be almost exactly expressed by this function of its attained horizontal distance at any *the same point*, x being that distance, viz.:

$$y = b\left(x - \frac{x^2}{a}\right),$$

in which a represents the whole horizontal distance attained, *with reference to the level of the point* of commencing to move, or rather of the *commencing calculation*, and in which b represents such a number of times, or is such a factor to one fourth of a, as equals or expresses the greatest height attained by the body if uninterrupted. But b may be fractional.

The demonstration of this formula belongs to Mechanics.

120. In the particular case when the body is thrown most favorably to attain the greatest range or horizontal distance for the force used, b becomes of the value 1, and may consequently be erased from the formula. For $\frac{a}{4}$ may be found to be the maximum value of

$$x - \frac{x^2}{a} = y,$$

and this value accrues to the function y, when $x = \frac{a}{2}$.

121. In the particular case, when the body is thrown perpendicularly upward, since $x = 0$ and $a = 0$, the general formula,

$$b\left(x - \frac{x^2}{a}\right) = y,$$

being restored, it would, at first thought, appear that y, or the height attained, must be zero, which would be absurd; but b being conditioned to be equal to *such* a number of times $\frac{a}{4}$ ($= 0$) as would make the actual greatest height attained, must be infinite, so that

$$\infty \times 0 = y,$$

where y may still have any finite value.

122. In the particular case, when the body is thrown horizontally, and attains no height after starting, each of a and b are zero, and the one only possible value of x is zero, with reference to *that line* of level. But a new line of level may be any where assumed, or the point where the body strikes or finds its course interrupted. Indeed, the descent of a thrown body is in its course, with reference to horizontality, symmetrical with its ascent, and we have shown elsewhere the symmetricalness of the function

$$50\,x - x^2 = y,$$

which, if $50 = a$ and $\frac{y}{50} = y'$, becomes $x - \frac{x^2}{a} = y'$; but it may have any factor or b, and the symmetricalness shown. Hence, the point of interrupted motion of a thrown body may be considered, for all the purposes of these calculations, as the point of projection.

123. Hence, a body thrown descendingly, describes the course of a body thrown upwardly, and having its upward motion interrupted before it may be completed; this being so, whether directly downward or laterally be the direction.

124. And hence further, a body allowed to fall directly down, is thrown by its gravity; and its point of setting out is the same as the point of greatest height that would be attained in the case of the same body thrown directly up, with the force such as it would have acquired in a fall through that same distance.

125. If the function given as expressing the variable height of the thrown body, be differentiated, we have

$$y = b\left(x - \frac{x^2}{a}\right),$$

$$\frac{dy}{dx} = b - \frac{2b}{a}x.$$

The value of this differential coefficient expresses the direction of motion of the body, while at the distance x on its course, with reference to horizontality and perpendicularity; it is their ratio, dy being upward, dx lateral. This coefficient is evidently greatest in practicalness when $x = 0$; therefore a body is making upward the most directly at its point of setting out. It is zero when $x = \frac{1}{2}a$; the body is then for an instant of time moving horizontally; maximum height is attained.

In the case when a body is thrown in the most favorable direction to attain range for the given force, since $b = 1$, this coefficient then $= 1$, consequently the body is thrown in the direction of the hypothenuse of a right-angled triangle, such as has its base and height equal, or at an angle of 45 degrees.

126. When an elastic thrown body strikes a firm perpendicular plane, as when a playing ball strikes the side of a building, the course and distance of the rebound are almost the complement of the course and distance as they would have been had there been no interruption.

127. Since the points of the setting out and of arrest of a thrown body in practice, are determined by human economy, or the presence and interruption of the earth, such points are special, and have no significance in the general mathematical indications of its course, which are more completely fulfilled in the case of moving celestial bodies. The thrown body may be conceived to have come out of the earth, and to again pass into or through it.

128. If, furthermore, we will discharge the condition that the successive lines of the measured heights of the thrown body are to be parallel, and will assume them to be parts of the radii of the earth, which is the more proper consideration, then will the thrown body return in its orbit into and out of the earth, or through it and back, and continue to revolve forever. But the formula would need some change to be rendered compatible with this course. Hence, further, a body thrown or dropped towards the centre of the earth's gravity, must, after going as far beyond, return in the same line, and oscillate forever, if uninterrupted.

129. If a body could be conceived as thrown with an infinite force, so that its range, a, may be also infinite, the formula

$$b\left(x - \frac{x^2}{a}\right) = y, \text{ becomes } b\,x = y,$$

because the fraction of which a is the denominator, becomes zero (unless x is also infinite, when which is the case, the fraction has an indeterminate value). The equation reduces to one of the first degree; its first differential coefficient is always b or constant; hence the body moves in a straight line forever.

A projected body always moves in a plane that is perpendicular to the horizon, or very nearly so.

1. (*a.*) A stone is so thrown as to reach the greatest distance, and its greatest height attained is 40 feet. Required the distance of its arrest or its range.

Ans. 160 feet.

(*b.*) Required the value of ratio of its upward and forward directions at its start. Ans. 1.

(*c.*) Required the same after it had attained the elevation of 16 feet, or while possessing that elevation.

In this case,

$$x - \frac{x^2}{160} = 16 \therefore x = 18 \text{ or } 142;$$

$$\therefore \frac{dy}{dx} = 1 - \frac{2 \times 18}{160} = \frac{31}{40},$$

$$\frac{dy}{dx} = 1 - \frac{2 \times 142}{160} = -\frac{31}{40}.$$

The former of these results is adapted to the ascension, the latter to the descension.

2. A stone is so thrown as to attain for its greatest height 62 feet, and distance 142 feet; it is required to determine how far it was from, when directly over or under, as it may be, the telegraph wire, which crosses the plane of the stone's motion, at 40 feet distance, horizontal, and height 50 feet.

3. A steam fire-engine threw water, the pipe being directed to a point 30 feet high at 70 feet distance; the force used carried the water a distance of 274 feet, measured on the level of the mouth of the pipe. Required the greatest height attained by the water above that level.

Ans. 29.35 feet.

4. From an elevation of 44 feet a body is thrown in its first outset downwardly 3 and laterally 5, *in direction of aim;* the force used was such that had it been exerted in precisely the opposite direction, the body would have risen to 17 feet greater height than that elevation. Required to determine how far from the foot of that elevation the body struck the ground. The direction of aim is supposed to be tangent to the course adopted by the body.

In this case we readily find the range the body would have made on the level of its setting out, had it been thrown in precisely the opposite direction, to be 108 feet $=a$; then with the variables x and y adapted to this level, we find $\frac{dy}{dx}$ for the value of $y=-44$, which value we will denote as,

$$\frac{dy'}{dx'}=b'=-\frac{142}{135},$$

to obtain which, $b=\frac{3}{5}$, was called negative. Adopting now the lower level as the basis of calculation, x' and y' as the variables, we have $17+44$, to find a', the lower range, from one half of which we subtract one half of a for the answer; the negative sign of $\frac{dy}{dx}$ is adapted to descent, and may be called positive with reference to the result desired, if the motion of the thrown body be construed as upward.

5. A person undertaking to draw a liquid from a small hole bored horizontally into the head of a barrel, observed that when the stream had obtained the downward descent of 7 inches, its lateral reach was 9 inches from the barrel's head, the same being supposed perpendicular. The hole was 27 inches above the floor, upon which stood a cylindric tin vessel of 4 inches diameter at the top and of 6 inches depth; the central axis of the vessel in the plane of the

stream was 18 inches laterally away from the perpendicular plane of the barrel's head. Required to determine whether the liquid was caught; the diameter of the stream being considered zero.

130. It is an important general principle to state, that the direction of the aim of an upwardly thrown body is always towards a *point just twice as high as the body ever reaches* at its greatest height, when such *point* is taken perpendicularly above the point of the greatest elevation that is to be actually attained without interruption. Hence, in the problem of the drawn liquid, the differential coefficient of the function descriptive of the course of the stream at the point 7 inches down, 9 inches lateral change, is $\frac{14}{9}$, if the direction of motion be considered reversed. But the most direct way of solving this problem is on the principle, true with reference to all falling bodies, when the point of commencing descent is given, that distances attained downward from this point are direct to each other as the squares of the laterally attained distance; hence,

$$7 : 21 :: 9^2 : Z^2,$$

Z being the lateral distance of stream when it is down 21 inches.

131. If the formula (Art. **119**), be considered with reference to the Degree of its Equation, it will be found to be of the Second, and to involve the Parabolic condition (**48**), because in

$$y = b\left(x - \frac{x^2}{a}\right),$$

i. e., $$a\,y + a\,b\,x - b\,x^2 = 0,$$

we have $a = D$, $a\,b = E$, $b = C$, $0 = B$, and $0 = A$; $\therefore B^2 - 4\,A\,C = 0$.

The change of the formula suggested in Art. **128**, but omitted, would give us an equation involving the Elliptical condition.

Gunshot projectiles being thrown at velocities varying from 900 to 2000 feet per second, the resistance of the air condensed before the moving body becomes so great that the formula given at (**119**) is unavailable; the body falls short of the distance indicated by the law of that formula.

SECTION XV.

THE SYSTEM OF SYMBOLS FOR FUNCTIONS. — AN IMPLICIT FUNCTION OF A SINGLE INDEPENDENT VARIABLE AND ITS DIFFERENTIATION.

132. We have shown the nature and purpose of a function of a single independent variable when it is explicit, and have exhibited the method of notation of it as (a function of x) $= y$. When more than (one function of x) $= y$ are employed in one investigation, it will be necessary to discriminate them; they are accordingly discriminated, as $F\,x$, $f\,x$, $f'\,x$, $f''\,x$, etc. Whenever we are concerned with two functions of x, which need not be alike, we should not use $F\,x$ as a notation for each of them.

133. We have also given in the first Section (Art. **11**) several equations which are generated from supposed conditions in a problem, in such a way that no function of x stands equated with a single equivalent in y. Such are *implicit* or *implied* (functions of x) $= y$. We are able, however, in simple cases of this kind to elaborate from the equation the explicit (function of x) $= y$. When we may not readily do this or cannot, we allude to the condition after transferring both members to one side of the equation (leaving, of course, 0 in the other) as a (function of x and y) $= 0$, or as $F(x, y) = 0$, and here it is understood that any and all terms consisting of isolated constants, are included in the designation; it is enough that x and y are somehow present. Of course in just this *notation*, viz., $F(x, y)$, we find no algebraic association of quantities; the whole expression is the equivalent of a sentence in language; the comma between x and y is intended to assist in dispelling any notion of a product of x by y, or of any specific algebraic association.

134. The general equation of the Second Degree (Art. **43**) shows the most involved relation possible for this Degree between x and y; A, B, C, etc., being supposed to be constants, this equation may be cited as $F(x, y) = 0$.

135. There is a peculiar significance in $F(x, y)$ being equated with 0; it could not be equated in general notation with any thing else, for every case of such use. Now in $F(x, y) = 0$ there is still the independent variable x, and the dependent variable y, as ever; they are mutual dependents, intended as ever preserving that combination of themselves with each other, and perhaps with constants equal to zero.

136. But if x and y each stand for quantities which are to vary independently, the value 0, which in any such particular case cannot vary, should be replaced, say with z, to denote the corresponding values, so that we shall have

$$F(x, y) = z.$$

137. An extension of this notation to an implicit function of two independent variables, gives us

$$F(x, y, z) = 0.$$

138. If there be three independent variables, the notation would be

$$F(x, y, z) = w,$$

and so on. This is the system of symbols for the general designation of implicit functions of variables.

139. In algebraic *equations with two or more unknown quantities*, in which conditions are furnished for the notation of the calculus, viz.:

$$F x + f y + f' z = a,$$

$$f'' x + f''' y + f'''' z = b,$$

$$\text{etc.,} \qquad \text{etc.,} \qquad = c,$$

means are supposed to be furnished by the independence of these equations, for the elimination successively, say of all functions of z, then of y, till we derive some (function of x) = some combination of a, b, c, etc., thence to find x, thence y, thence z. But such functions must be of the simplest form to render this possible.

140. We proceed to show that it is not necessary to deduce in form an explicit function of an independent variable from the implicit state, that we should be able to differentiate such explicit function as to its dependent variable; that is, we may find $\frac{dy}{dx}$ from $F(x, y) = 0$ without finding $f x = y$.

141. It will have been understood in differentiating a case of $F x = y$, and making the notation $d F x = d y$, the terms of an actual particular function taking the place of $F x$, that we have given two names for the same amount. The function is one. We pass the sign of equality, and express by $d y$ what is also expressed in the other member of the equation. This consideration would at once authorize us to transpose y previously, so that from

$$F x - y = 0,$$

we should have $\quad d F x - d y = 0.$

Again, there is obviously no reason but that of simplicity why we equate $F x$ with y rather than with $a y$ or y^2, i. e., with *some function* of y. If u be such a function of y in amount, we might have for differentiation of x with respect to y as a dependent variable, such a case as

$$b x^3 + x = y^2 - a y = u;$$

differentiating $f y = u$, we have

$$2 y d y - a d y = d u,$$

where y may be independent with respect to u, but by hypothesis y may be still also dependent with respect to x. Since $F x$ is equated with u, we are entitled in differentiating it with respect to u, to the following equation, and in doing this we do not recognize y as other than a constant:

$$d F x = 3 b x^2 d x + d x = d u,$$

so that we may infer that we are entitled to the following equation (remembering that the differential of $f y$ depends on y):

$$3 b x^2 d x + d x = 2 y d y - a d y = d u:$$

from which we are able to deduce an expression for $\frac{d y}{d x}$, as follows:

$$3 b x^2 + 1 = 2 y \frac{d y}{d x} - a \frac{d y}{d x} = \frac{d u}{d x};$$

$$\therefore \frac{3 b x^2 + 1}{2 y - a} = \frac{d y}{d x}, \; = \frac{d u}{d x} \div (2 y - a), \; = \frac{d u}{d x} \div \frac{d u}{d y}.$$

142. Now, if instead of the above case of $F x = f y$, we had had the same in the form of $f' (x, y) = 0 = u'$, that is,

$$b x^3 - y^2 + x + a y = 0 = u',$$

there occurs nothing that would alter the result for $\frac{d y}{d x}$. Here we have given an accent to u for denoting that u' is used for $f' (x, y)$ instead of for $f y$, as before. But why do we use u' at all since we have zero also, and since $d u'$ is manifestly 0? The answer is, $u' = 0$ and $d u' = 0$ always; but we are proceeding to make the fraction $\frac{0}{0}$, and instead of $\frac{0}{d x}$ or $\frac{0}{d y}$ we find it more useful to use something that will preserve a special relation to $f' (x, y)$, and be a means of discrimination whenever we should have associated in an investigation implicit functions which we might be obliged to discriminate as $f (x, y) = 0 = u$, and $f (z, w) = 0 = v$, etc.

Zero does not, in the ordinary use as 0, preserve its record and identity; does not take account of its factors; if $f (x, y) = 0$, then $10 \times f (x, y) = 0$, but in the use of u we should have $10 u$.

It is an important principle in differentiation to observe, that we allow no loss of quantities, such as the factors of zero, and receive none such of which we do not make a record.

We may generally place $(f (x, y) = 0) = u$, for the purpose of differentiation, without the accent.

Let the next be a more intimately connected case of $f (x, y) = 0$, viz.:

$$a x y^2 + b x^2 y + x^2 + y^2 + c = 0 = u,$$

$$\therefore 2 a x y d y + a y^2 d x + b x^2 d y + 2 b y x d x + 2 x d x + 2 y d y = 0 = d u;$$

$$\therefore 2\,a\,x\,y\frac{d\,y}{d\,x}+a\,y^2+b\,x^2\frac{d\,y}{d\,x}+2\,b\,y\,x+2\,x+2\,y\frac{d\,y}{d\,x}=$$
$$0=\frac{d\,u}{d\,x};$$

now, $$2\,a\,x\,y+b\,x^2+2\,y=\frac{d\,u}{d\,y};$$

$$\therefore -\frac{a\,y^2+2\,b\,y\,x+2\,x}{2\,a\,x\,y+b\,x^2+2\,y}=\frac{d\,y}{d\,x}=-\frac{d\,u}{d\,x}\div\frac{d\,u}{d\,y}.$$

In finding $\frac{d\,u}{d\,y}$ we differentiate u with respect to just the variable y.

If the above particular demonstration would authorize a general rule, such rule would be as follows: but a general demonstration would be quite abstruse. While, then, the rule is merely stated, it is recommended, for the purposes of the following Section, to work out the results for $\frac{d\,y}{d\,x}$ in a manner similar to the above, by differentiating terms in $F(x, y)=0$, when the following rule will be found to be a declaration of each result:

143. Whenever we have a case that can be cited as $f(x, y)=0$, and we wish to derive $\frac{d\,y}{d\,x}$, *we differentiate the expression as if y were a constant, and divide the coefficient so obtained by the coefficient obtained from differentiating the same expression on the supposition that x is constant, and then change the sign of the fraction so obtained; this fraction is* $\frac{d\,y}{d\,x}$.

Problems.

1. In the implicit function of the variable x in $x^2+10\,y+5\,x=0$, required $\frac{d\,y}{d\,x}$. Ans. $\frac{d\,y}{d\,x}=-\frac{2\,x+5}{10}$.

2. Required $\frac{dy}{dx}$ in $x^2 + y^2 - 50 = 0$.

$$\text{Ans. } \frac{dy}{dx} = \quad \frac{x}{y} = \pm \frac{x}{\sqrt{50 - x^2}}.$$

This case shows an obvious reduction of an implicit dif. coef. reduced to an explicit one; the previous case did not require such reduction. In general, the explicit dif. coef. may not be easily found, or may not be needed; all the purposes of a dif. coef. being subserved by the implicit dif. coef.

3. Required $\frac{dy}{dx}$ in $x^2 + y^2 - m x y - 81 = 0$.

$$\text{Ans. } \frac{dy}{dx} = \frac{m y - 2 x}{2 y - m x}.$$

4. Required $\frac{dy}{dx}$ in $x^3 - 3 c x y + y^3 = 0$.

$$\text{Ans. } \frac{dy}{dx} = \frac{c y - x^2}{y^2 - c x}.$$

5. Required $\frac{dy}{dx}$ in $(x - a)^2 + (y - b)^2 = 0$.

$$\text{Ans. } \frac{dy}{dx} = \frac{x - a}{y - b}.$$

6. Required $\frac{dy}{dx}$ in $24 x^2 y - y^3 + 10 x = 0$.

$$\text{Ans. } \frac{dy}{dx} = - \frac{48 x y + 10}{24 x^2 - 3 y}.$$

144. It is scarcely necessary to say the successive differentiation of $f(x, y) = 0$, relatively to y as dependent and x as independent variable, may be performed; and that Taylor's Theorem and the theory of maxima and minima are available for $F x = y$ in the cases given as $f(x, y) = 0$.

145. But $f(x, y) = 0$ is constant, and can have no

maximum or minimum, as is very obvious, and in making this remark we are not alluding to $f' x$ involved in $f(x, y) = 0$.

146. In a treatise of algebra the following are given as illustrations of two equations that are *contradictory*, because the same value for x is not deducible from each:

1. $3x = 60$;

2. $\frac{1}{2} x = 20$.

But suppose we proceed to consider them in this way:

3. $3x - 60 = 0'$;

4. $\frac{1}{2} x - 20 = 0''$;

5. $\therefore 3x - 60 = \frac{1}{2} x - 20 \therefore x = 16$;

whence the question arises, what is there illogical about such a course of proceeding? although the result is not compatible with $x = 20$ or $x = 40$, as would be severally deduced from (1.) and (2.). The answer is, that x in (3.) is an alien from x in (4.), and by *independence* it renders $F x = 0$ in (3.), and $f x = 0''$ in (4.); or x, for the instant considered a variable, produces the value $0'$ in (3.), and $0''$ in (4.), by an independent law of change. In regard to equation (5.), we can say that the value $x = 16$, is such as will truly render $F x = f x$. Now, $\frac{d F x}{d x} = \frac{6 d f x}{d x}$, the functions do not change at like rates. Indeed, the zero $0'$ in (3.) and $0''$ in (4.) are not produced from like elements, and are not compatible with each other, and *not equal.* No better proof could be desired of the statements made in previous Sections in regard to the diverse values of zero, as dependent on distinct origins.

Suppose we have the two algebraic equations,

1. $3x = 0', \therefore x = 0'''$;

2. $\frac{1}{2}x = 0'', \therefore x = 0''''$;

whence if $0' = 0''$, we have

3. $3x = \frac{1}{2}x$;

dividing by x, $3 = \frac{1}{2}$;

or, $3x - \frac{1}{2}x = 0' - 0''$;

$$\therefore x = \frac{0' - 0''}{3 - \frac{1}{2}} = 0'''''.$$

These results are only to be reconciled by the diverse values of zero, as well as of x when at zero.

Suppose next we have the two equations,

1. $3x - 60 = 0 \therefore x = 20$;

2. $3x = 0' \therefore x = 0''$;

whence if $0 = 0'$, we have

3. $3x - 60 = 3x$;

$$\therefore x = \frac{60}{0} = \infty.$$

Now, in no case are the equations (3.) absurd in nature, but the mode of making them is not consistent with (1.) and (2.) in the three supposed cases.

147. In a former part of this section, the nature and the differentiation of a (function of x and y) $= 0$ were discussed: there might have been suggested, in that connection, the question, if we have two functions of x, Fx and fx, which might be equal, whether, when they are, their differentials are also *necessarily* equal? But they are not. This being contrary to what a superficial view would lead us to adopt, is worthy of a statement of the reasons.

In the simple condition given of $F x = f x$, in the maintenance of which x in $F x$ has a particular value, it is pure presumption to suppose that this is the same as that of x in $f x$, at which x in $f x$ gives $f x$ the same value.

In the case of $F x = y$, we call x an *independent* variable; also in the case of $f x = y'$, we call this x an *independent* variable. Now in the accommodation of the *particular* condition of $y = y'$, x in $F x$ and x in $f x$ must retain their independence as ever, as more likely to accommodate the condition, which a simple example should show. Let $F x$ be $x^2 - 40 x + 1802$, and $f x$ be $80 x - x^2$, then if $F x = f x$, we have

$$x^2 - 40 x + 1802 = 80 x - x^2;$$

but we *cannot deduce any common value* of x in $F x$ and $f x$, by which this possible equation is sustained.

When we come to the dif. coefs. of $F x$ and $f x$, they cannot be inferred to be necessarily equal for the mere reason that the functions from which they have been derived happened to be such that they might have a common value, but nothing else of *nature* in common. Differential coefficients show the *nature* of functions through all values.

The exhibition of many cases of functions of x and their being equated with y, tends to create an illusion as to their entire incompatibility when they enter into random association as above. They should at once be expressed in proper language for such association, as $F x = y$, $f z = v$, $f' w = u$, etc.

SECTION XVI.

PROBLEMS WHICH MAY FURNISH IMPLICIT FUNCTIONS OF ONE VARIABLE, AND CASES OF THEIR MAXIMA AND MINIMA.

1. A drover bartered 160 head of cattle for sheep: the number of sheep obtained in the exchange was found to be 40 times the dollars allowed for the value of a sheep. It is required to determine the law or rule by which the value of one of the cattle, in number of dollars, will vary in the fulfilment of these conditions, compared with the varying value of a sheep.

Let $y =$ number dollars for 1 of the cattle,

and $x =$ number dollars for 1 of the sheep;

then $40\,x^2 = 160\,y$, or $40\,x^2 - 160\,y = 0$;

$$80\,x\,dx = 160\,d\,y;$$

$$\therefore \frac{d\,y}{d\,x} = \tfrac{1}{2}\,x.$$

Ans. The number of dollars value of one of the cattle, tends to increase as many times, or as much of a time, the number of dollars value of a sheep, at any supposition for either, as $\frac{1}{2}$ of what the number of dollars value of a sheep, at any such supposition, may be.

2. A fruit-seller carried a certain number of bushels of fruit to market, which he sold, and expended $8.50 of the proceeds for grain, when he found that he had in money, unspent, the value of 5 bushels of the fruit. If there had been 22 bushels of the fruit, by supposition, and then just

an increase in that number be suggested, what effect will this suggestion have on their change in value per bushel, in still fulfilling the conditions of the problem?

Let $x =$ the number of bushels of fruit,

and $y =$ the number of cents per bushel;

$$\therefore x\,y - 850 = 5\,y;$$

$$\therefore y\,d\,x + x\,d\,y = 5\,d\,y;$$

$$\therefore \frac{d\,y}{d\,x} = -\frac{850}{(x-5)^2} = -\frac{850}{289}.$$

Hence the price in cents must *commence* to diminish at the rate of $2\frac{252}{289}$ times the number of bushels should be supposed to increase. This incipient ratio is expressed in these units, but would itself vary *during* the change of as much as the *whole* unit, one bushel.

3. (a.) A courier rode 30 hours in all, but successively on a gray and a red horse; the number of miles on the gray one was 7 times the miles he went per hour on the red one; the number of miles on the red one was 9 times the miles he went per hour on the gray one: it is required to determine how the miles per hour on the gray one must be inferred to change relatively to the number per hour on the red one, on any suggestion of change of number of miles per hour on the red one, when either is at any possible rate allowed by the conditions.

Let $x =$ miles per hour on red horse,

and $y =$ miles per hour on gray horse;

$$\therefore \frac{a\,x}{y} = \text{number hours on red horse};$$

and $\frac{b\,y}{x} =$ number hours on gray horse;

$$\therefore \frac{7x}{y} + \frac{9y}{x} - 30 = 0;$$

$$\therefore \frac{dy}{dx} = -\frac{1}{2 \times 9}(30 \pm \sqrt{30^2 - 4.7.9}) = -3 - \tfrac{1}{4} \text{ or } \tfrac{8}{100};$$

$$\therefore \frac{dx}{dy} = -4 \text{ or } -\tfrac{32}{100};$$

$$\therefore \frac{d^2y}{dx^2} = 0 \text{ and } \frac{d^2x}{dy^2} = 0.$$

Whence it appears that y, at whatever value it may have, will increase minus $\frac{1}{4}$th, or minus 3.08 times as fast as x. Since $\frac{dy}{dx}$ is constant, and cannot $= 0$, there is no limit to y. Because $\frac{dx}{dy}$ is constant, there is no limit to greatness of x. Since in

$$7x^2 + 9y^2 - 30xy = 0,$$

if $y = 0$, $x = 0$, there is not necessarily any distance attained by the courier, whether he sits on the horse's back 30 hours, or any other number of hours, and the word *ride* seems to fail in practicability in that condition. But his rate of motion may be indeterminately great.

(*b*.) A courier rode 30 hours in all, but successively on a gray and a red horse; the number of miles on the gray one was 7 times the miles he went per hour on the red one, *and* 1 *mile more ;* the number of miles per hour on the red one was 9 times the miles per hour on the gray one: it is required to determine how the miles per hour on the gray one must be inferred to change relatively to the number per hour on the red one, on any suggestion of change of number of miles per hour on the red one, when either is at any possible rate allowed by the conditions, and whether there are maxima of miles per hour on each horse.

It will be ascertained that the words in this problem, "*and* 1 *mile more*," change, very materially, all the characteristics of the results of the solution, in comparison with problem (*a.*).

4. (*a.*) A caterer having 9 dollars in gold and a number of dollars in silver, purchased at a market 90 quails and as many more quails as he had dollars in silver, of such value each quail, that his silver alone would have paid for 144, and then invested the balance of his money in as many pigeons as quails already bought. Required the number of dollars in silver if his quails cost the greatest possible sum each; and required that sum.

Ans. 18 dollars, and 12½ cents each pigeon.

(*b.*) Required what two several numbers of dollars he may have had in silver that the quails alone should have exhausted just all his money.

The conditions of the above problem (*a.*) may be compared with those of Problem 27, in Section XII.

5. A man purchased a commodity known as *A*, for which he paid as many cents a pound as it weighed pounds, and the number of pounds of *A* differed by 10 from the number of pounds of the commodity known as *B*, which was his second purchase. The next day he purchased the commodity distinguished as *C*, at the same number of cents a pound as there were pounds, and its number of pounds differed by 15 from that of the number of pounds of the commodity known as *D*, his fourth and last commodity purchased. But we know of no guarantee that each of the commodities *A* and *C* had weight, though it may be inferred that one of them must have had. He paid for *A* and *C* 81 cents. It is required to determine the possible range of relative weights of *B* and *D*, and the

greatest and least weights of each of these two commodities.

Ans. Limits of B, 1 and 19 pounds; of D, 6 and 24 pounds.

6. (*a.*) A furrier purchased two lots of furs: when asked what he paid per pound for the kind A, he said, this purse in my hand contains 99 dollars in coins; if I take out as much value as I paid per pound for the kind B, and repeat this as many times as I take out dollars each time, and if I then take the money left in the purse and divide it into as many piles as I put dollars in a pile, one of these piles is the price I paid. When asked how many pounds he purchased of A, he said $2\frac{5}{9}$ pounds more than I just took out dollars out of the purse at once, and the amount of money I expended for each lot of fur was alike. Required a limit of the price of B.

(*b.*) If we adopt the principle of assuming a quantity for B, with the intention afterwards of exchanging it for a quantity the least greater, at what amount of B are we stopping in the consideration, when the suggested increase involves the most rapid decrease of its price per pound.

7. (*a.*) A counterfeiter, escaping arrest, rode a black horse as many hours as miles per hour, when he exchanged and rode a white horse as many hours as miles per hour. At the instant he commenced riding the white horse, an officer commenced a pursuit from the original point, at six fifths of the speed of the black horse. When the fugitive is done with the white horse, he is just 81 miles from the officer, at which point he is arrested through the aid of the telegraph. The owners of the two horses which had been pressed into the above service by the fugitive, on recovering them, were anxious to know which animal had been hardest and longest driven. It is required to determine that particular condition, according to which either horse's service was the greatest the conditions admit of.

Let $x =$ black horse's hours and miles per hour,

and $y =$ white horse's hours and miles per hour,

and $c =$ six tenths;

then $2\,c\,x\,y =$ the officer's travel in miles;

then $x^2 + y^2 - 2\,c\,x\,y - 81 = 0;$ (1.).

$$\therefore \frac{d\,y}{d\,x} = \frac{c\,y - x}{y - c\,x} = 0 \text{ when } y = \text{max.};$$

$$\therefore c\,y - x = 0 \text{ and } y = \frac{5}{3}\,x;$$

substituting this value of y, for y in (1.),

$$x = +\ 6\tfrac{3}{4} \text{ or } -\ 6\tfrac{3}{4},$$

$$\therefore y = +\,11\tfrac{1}{4} \text{ or } -\,11\tfrac{1}{4}.$$

Discarding the negative values of x as impracticable, for an event cannot take place through negative time, we wish to know whether the positive value of x gives a maximum or minimum for y, by examining the $\frac{d^2\,y}{d\,x^2}$;

$$\frac{d^2\,y}{d\,x} = \frac{(y - c\,x)\,(c\,d\,y - d\,x) - (c\,y - x)\,(d\,y - c\,dx)}{(y - c\,x)^2};$$

$$\frac{d^2\,y}{d\,x^2} = \frac{c\,y\frac{d\,y}{d\,x} - c^2\,x\frac{d\,y}{d\,x} - y + c\,x - c\,y\frac{d\,y}{d\,x} + x\frac{d\,y}{d\,x} + c^2\,y - c\,x}{(y - c\,x)^2} =$$

$$\frac{(c^2\,y - y)}{(y - c\,x)^2} \text{ since } \frac{d\,y}{d\,x} = 0;$$

$$\therefore \frac{d^2\,y}{d\,x^2} < 0 \therefore y \text{ when } 11\tfrac{1}{4} = \text{maximum.}$$

It is to be remembered that $\frac{d^2\,y}{d\,x^2}$ in the *general* state is the *full* one foregoing; for $\frac{dy}{dx}$ has been made special when made $= 0$.

(*b.*) How many miles per hour do the conditions require the white horse to travel when the black one travels 9?

Ans. $10\frac{4}{5}$ or none.

(*c.*) How many miles per hour do the conditions require the white horse to travel when the black one travels 12?

Ans. The supposition is forbidden; so great a value for x as 12 is imaginary.

(*d.*) What two values satisfy the conditions for the white horse's speed if the black one's is made 10 miles an hour. Ans. $10\frac{3}{25}$ and $1\frac{17}{20}$ very nearly.

It may be interesting to notice that the maximum speed and duration of service of either horse is not that which most favored the other, because the distance executed by the fugitive is not an absolutely determinate amount. The greatest distance attained by the officer is evidently when $x = y$ in $2\,c\,x\,y$, which occurs when $x = y = 10.035$ miles or hours; and since the fugitive went 81 miles more, the value of x and $y = 10.035$ gives the greatest associated service of each horse.

8. Tradition says that a certain king erected a solid structure of stone 20 (a) rods long, its width and height alike; that his successor, or the second king, erected two structures, the one 12 (b) rods long, the height and width alike, the other 12 (b) rods long, 8 (c) rods wide, and as high as the one last mentioned. His successor, the third king, converted all of the three previous structures into two of his own, the one a complete cube, the other 8 (c) rods wide, and of that uniform height of all the structures of the second and third kings, and as long as high, it being understood that the mere names, *width* and *length*, are interchangeable when necessary.

The question is the height of the first king's structure, what it might have been, and whether it existed at all, if the second king built his less than 12 rods high.

The negative values of any quantity being impracticable within this problem, the study of the function at negative values of the variables bring out very peculiar properties, which are best shown by a diagram.

9. (*a.*) For a certain purpose two cubical cisterns, A and B, are needed, and one a rectangular prism, C. The cubical cisterns, A and B, are to have, combined, the same capacity as C. C must have one linear dimension 15 feet, another the same as that of cistern A, the other the same as the cistern B. A well-informed contractor agrees to make the cistern B for a stipulated sum. The other party, designing to receive the largest possible cistern for the fixed sum, studies to vary the size of the cistern A. Required to know all the dimensions of the three cisterns when the one here contracted for is the largest possible.

Ans. A, $5 \times 2^{\frac{1}{3}} = 6.299$ feet.
B, $5 \times 4^{\frac{1}{3}} = 8.025$ feet.

(*b.*) Since either of the cubic cisterns can have a linear dimension 8.025 feet, if one be made only 7 feet, required the linear dimensions of the others.

(*c.*) If the two cubic ones are equal, how large are they?

(*d.*) Do all the cisterns become of no capacity if either one does?

10. (*a.*) A balloonist being asked to give some account of the heights and distances of his latest ascension, said that y, conditioned as follows, was his height in miles every where on his voyage over the level country, while x was his distance by horizontal measure in miles from his point of beginning to rise, viz.:

$$48\,y = 192\,x - 88\,x^2 + 16\,x^3 - x^4;$$

in which we have $F y = f x$, and implicitly $y = f' x$, and in this simple nature of $F y$ it is evident we may easily express $y = f' x$, which is the function of x that is significant in the problem. Required the prominent points in the history of heights and distances of the voyage, such as:

(*b*.) What was the distance made at the landing?

(*c*.) What and where was his greatest height, if there was any greater than all others?

(*d*.) To determine if, after first descending any, he afterwards ascended, and where, and being how high he commenced any second ascent?

(*e*.) To determine if, in rising the second time, he went up to the same greatest height to which he had ascended?

(*f*.) How far apart were any two places at which he began to descend, measured on the ground line?

(*g*.) Where in the voyage he rose most directly upward?

(*h*.) How far from the point of starting, ground measure, was he at each of his greater heights?

(*i*.) If he had gone in a straight line in the direction he first started, how high would he have been when 200 feet from his starting point, ground measure?

(*j*.) Had he risen out of the earth and descended into it after alighting, would the function indicate the law of his course beneath the earth's surface?

(*k*.) Specify the term in the function which intimates that he must rise. Ans. $192\,x$.

(*l*.) Specify the term which intimates that he must finally come down. Ans. $-x^4$.

(*m*.) Specify the term which intimates the *probable* originating of a second place of rising; probable, because, in the generalization of the constants, such term might be neutralized by another one just equal to it in value, with an opposite sign.

(*n*.) To what distance below the earth's surface does the

function indicate the law of his course, including the consideration of x when negative?

(*o.*) Is there any thing symmetrical between the first and last halves of the voyage?

(*p.*) If the sign of every term in the function be changed, is the voyage indicated as a dive below the earth's surface, and a final emergence from it?

(*q.*) Then what are the presumptions about the course before and after diving?

The first dif. coef. is of the third degree, but may be resolved by trials with integral numbers for x.

The presumption has been that this aerial voyage was performed in one perpendicular *plane;* but it will be perceived not to be essential if the ground line, however tortuous, is considered to be beneath the voyage track, and measurable like a straight line.

(*r.*) It is required to alter the ascertained $y = f'x$ (perhaps by factor common to every term), so that the greatest heights may be expressed as at $3\frac{1}{2}$ miles, and all other heights in proportion.

(*s.*) Need such a change alter the distance of the landing?

(*t.*) It is required to modify $y = f'x$ so that the place of the landing may be 15 miles, without affecting the heights as first conditioned.

(*u.*) It is required so to give out the Fy that the place of landing may be 14 miles, and greatest heights 2 miles in the same connection.

11. (*a.*) There is to be determined the size of a square piece of land, which it is proposed to enclose with a fence, at the cost of three or a times as many dollars per rod in length, as is the worth of as many square rods of the land as each of said square rods is worth dollars; and the cost

of the whole fence will be 60 or b dollars less than the the worth of the land, and the land the smallest possible.

Let $y =$ the number of square rods,

and $x =$ the number of dollars per square rod;

then $4\,y^{\frac{1}{2}} =$ the number of linear rods round;

and $4\,a\,y^{\frac{1}{2}}\,x^2 =$ the dollars cost whole fence;

then $$4\,a\,y^{\frac{1}{2}}\,x^2 - x\,y + b = 0, \qquad (1.).$$

and $$\frac{d\,y}{d\,x} = \frac{8\,a\,y^{\frac{1}{2}}\,x - y}{x - 2\,a\,x^2\,y^{-\frac{1}{2}}};$$

the numerator being $= 0$ in case $y =$ max. or min.,

$$\therefore y = 64\,a^2\,x^2,$$

by substitution in (1.)

$$x = \left(\frac{b}{32\,a^2}\right)^{\frac{1}{3}} = .5928;$$

and $$y = 64\,a^2\left(\frac{b}{32\,a^2}\right)^{\frac{2}{3}} = 64\left(\frac{a\,b}{32}\right)^{\frac{2}{3}} = 202.44;$$

$$\therefore \text{one side} = 13.974 \text{ rods.}$$

But to ascertain beyond doubt whether we have certainly either a maximum or minimum, and which, we must determine whether $\frac{d^2\,y}{d\,x^2}$ is zero, or positive or negative; we had:

$$\frac{d\,y}{d\,x} = \frac{8\,a\,y^{\frac{1}{2}}\,x - y}{x - 2\,a\,x^2\,y^{-\frac{1}{2}}};$$

$$\therefore \frac{d^2\,y}{d\,x} = \frac{(x - 2\,a\,x^2\,y^{-\frac{1}{2}}) \times (8\,a\,y^{\frac{1}{2}}\,d\,x + 4\,a\,x\,y^{-\frac{1}{2}}\,d\,y - d\,y)}{(x - 2\,a\,x^2\,y^{-\frac{1}{2}})^2} -$$

$$\frac{(-8\,a\,y^{\frac{1}{2}}\,x + y) \times (d\,x - 4\,a\,y^{-\frac{1}{2}}\,x\,d\,x - a\,x^2\,y^{-\frac{3}{2}}\,d\,y)}{(x - 2\,a\,x^2\,y^{-\frac{1}{2}})^2};$$

multiplying terms and dividing by $d\,x$, we have,

$$\therefore \frac{d^2\,y}{d\,x^2} = \frac{16\,a^2\,x^2 + y - 4\,a\,y^{\frac{1}{2}}\,x}{(x - 2\,a\,x^2\,y^{-\frac{1}{2}})^2}, \text{ when } \frac{d\,y}{d\,x} = 0,$$

the numerator of which $= \dfrac{48\,a^2\,b^{\frac{2}{3}}}{(32\,a^2)^{\frac{2}{3}}}$;

and therefore second dif. coef. is positive.

$\therefore y$ is found at a minimum.

$\therefore x$ and y are determinate by the concurrent equations;

$$4\,a\,y^{\frac{1}{2}}\,x^2 - x\,y + b = 0,$$

$$8\,a\,y^{\frac{1}{2}}\,x - y = 0.$$

NOTE. In consequence of the length of the above expression for $\frac{d^2 y}{d x}$, we have expressed the same in two terms having a common denominator; this accounts for the signs of $-8\,a\,y^{\frac{1}{2}}\,x + y$.

(*b*.) Required the size of the lot when the least number of dollars is paid per square rod for it, and what that number of dollars would be.

SECTION XVII.

FUNCTIONS OF TWO INDEPENDENT VARIABLES: THEIR DIFFERENTIATION AND THEIR MAXIMA AND MINIMA.

148. The mode of notation by which we may cite a function of *two* independent variables has been (Art. **136**) shown to be $f(x, y) = z$. From the circumstance of identity we must, therefore, have $d f(x, y) = d z$. Since z may vary on a variation of x; and since at the same time y need not vary on account of its independence, we express this variation of z with respect to x, by *differential coefficient*, as $\frac{d z}{d y}$; and the variation of z with respect to y by differential coefficient, as $\frac{d z}{d x}$.

There is, then, no better way for expressing, beyond doubt, the *whole differential* of z with respect to x and to y, by general notation, than by $(d z)$, or

$$\frac{d z}{d x} d x + \frac{d z}{d y} d y;$$

which becomes very intelligible if we remember that in a common case of $f x = y$, $d f x$ might have been cited as $\frac{d y}{d x} d x$, but which was unnecessary. In all particular cases, however, in this section, we shall have use for the expression of only the dif. coefs. $\frac{d z}{d x}$ and $\frac{d z}{d y}$. Although we express the whole differential coefficient of z as $\frac{d z}{d x} + \frac{d z}{d y}$, we use the signs in the general sense; particular conditions may render either, or both, negative in value, although the amount of the change of value of x and of y may be positive.

149. *In differentiating a (function of x, y) $= z$, we may conveniently express the dif. coefs.* $\frac{d z}{d x}$ *and* $\frac{d z}{d y}$ *in successive equations, that variable being considered a constant with reference to which we are not differentiating the equation. Their algebraic sum is the total differential coefficient required.*

1. Required the whole dif. coef. of $3x^2 y + x = z$.

Ans. $\frac{dz}{dx} + \frac{dz}{dy} = 6xy + 1 + 3x^2$.

2. Required the whole dif. coef. of $\frac{x}{y} = z$.

Ans. $\frac{dz}{dx} + \frac{dz}{dy} = \frac{y - x}{y^2}$.

3. Required (dz), or whole differential of $\frac{y}{3y^2 - x} = z$.

Ans. $\frac{dz}{dx} dx + \frac{dz}{dy} dy = (dz) = \frac{y\,dx - 3y^2\,dy - x\,dy}{(3y^2 - x)^2}$.

4. Required the whole dif. coef. of $15701\,xy + ax^2 = z$.

150. It is useful and quite important to extend Taylor's Theorem to embracing a development of a function of two independent variables, the condition being that each variable may concurrently take an increment or decrement, or one variable an increment and the other a decrement; which condition must embrace the case of variation limited to one of the variables.

151. In the function $f(x, y) = z$, if x take the increment h, the function will become $f(x + h, y)$, y remaining unchanged, since it is independent of x: then, by Taylor's Theorem,

$$f(x + h, y) = z + \frac{dz}{dx} h + \frac{d^2 z}{dx^2} \cdot \frac{h^2}{1 \cdot 2} +, \text{ etc.} \qquad (1.).$$

But if y also take an increment k, then z in the above expression becomes changed to

$$z + \frac{dz}{dy.} k + \frac{d^2 z}{dy^2} \cdot \frac{k^2}{1 \cdot 2} + \frac{d^3 z}{dy^3} \cdot \frac{k^3}{1 \cdot 2 \cdot 3} +, \text{ etc.,} \quad (2.).$$

so that in place of z in (1.), we must substitute *the whole* of (2.), and in doing so after we have passed z, we must for $\frac{dz}{dx}$ put, in (1.), the dif. coefs. (with respect to x), of every term in (2.); that is, we must substitute

$$\frac{dz}{dx} + \frac{d \cdot \frac{dz}{dy}}{dx} k + \frac{d \cdot \frac{d^2 z}{dy^2}}{dx} \cdot \frac{k^2}{1 \cdot 2} +, \text{ etc., for } \frac{dz}{dx},$$

$$\frac{d^2 z}{dx^2} + \frac{d^2 \cdot \frac{dz}{dy}}{dx^2} k + \frac{d^2 \cdot \frac{d^2 z}{dy^2}}{dx^2} \cdot \frac{k^2}{1 \cdot 2} +, \text{ etc., for } \frac{d^2 z}{dx^2},$$

$$\frac{d^3 z}{dx^3} + \frac{d^3 \cdot \frac{dz}{dy}}{dx^3} k + \frac{d^3 \cdot \frac{d^2 z}{dy^2}}{dx^3} \cdot \frac{k^2}{1 \cdot 2} \text{ for } \frac{d^3 z}{dx^3},$$

and so on. Before, however, making these substitutions for convenience only, and not as an algebraic act, let us agree to write

$$\frac{d^2 z}{dy\,dx} \text{ for } \frac{d \cdot \frac{dz}{dy}}{dx}, \quad \frac{d^3 z}{dy^2\,dx} \text{ for } \frac{d \cdot \frac{d^2 z}{dy^2}}{dx},$$

and, generally, $\frac{d^{q+p} z}{dy^q\,dx^p}$ for $\frac{d^p \cdot \frac{d^q z}{dy^q}}{dx^p}$.

152. Hence the result of the proposed substitutions in (1.), will give us

$$f(x+h, y+k) = z +$$

$$\frac{dz}{dx} h + \frac{d^2 z}{dx^2} \cdot \frac{h^2}{1 \cdot 2} + \frac{d^3 z}{dx^3} \cdot \frac{h^3}{1 \cdot 2 \cdot 3} +, \text{ etc.};$$

$$\frac{dz}{dy} k + \frac{d^2 z}{dy\,dx} kh + \frac{d^3 z}{dy\,dx^2} \cdot \frac{kh^2}{1 \cdot 2} +, \text{ etc.};$$

$$\frac{d^2 z}{dy^2} \cdot \frac{k^2}{1 \cdot 2} + \frac{d^3 z}{dy^2\,dx} \cdot \frac{k^2 h}{1 \cdot 2} +, \text{ etc.};$$

$$\frac{d^3 z}{dy^3} \cdot \frac{k^3}{1 \cdot 2 \cdot 3} +, \text{ etc.}$$

$$+, \text{ etc.}$$

The foregoing is the development, required by Taylor's Theorem, of a function of two independent variables, on each of them undergoing a change of value.

153. If every term containing k as a factor, disappear, as when k should be zero, then the development reverts to one for a single variable h.

154. If, however, h and k be negative, all those terms in the foregoing development where h or k occur at even powers, will evidently be positive, the others where h or k stands without the other will be negative; but since the development is supposed to hold also for

$$F(x - h,\ y + k),$$

or

$$F(x + h,\ y - k),$$

the terms in which h and k are factors together must be *ambiguous*, in the development, for it must be doubtful whether $h\,k$ arises from $-h \times -k$ or from $+h \times +k$.

155. Whenever z, in a case of $f(x, y) = z$, is at a maximum, we must have the condition

$$f(x, y) > f(x \pm h, y \pm k),$$

and consequently

$$\left(\pm \frac{d\,z}{d\,x} h \pm \frac{d\,z}{d\,y} k\right) + \tfrac{1}{2}\left(\frac{d^2\,z}{d\,x^2} h^2 \pm 2\,\frac{d^2\,z}{d\,x\,d\,y} h\,k + \frac{d^2\,z}{d\,y^2} k^2\right) + \text{etc.} < 0.$$

Now, since there is no reason why h and k in the above expression may not be alike, or each h, the above inequation may be written

$$\left(\pm \frac{d\,z}{d\,x} \pm \frac{d\,z}{d\,y}\right) h + \tfrac{1}{2}\left(\frac{d^2\,z}{d\,x^2} \pm 2\,\frac{d^2\,z}{d\,x\,d\,y} + \frac{d^2\,z}{d\,y^2}\right) h^2 +, \text{etc.}, < 0;$$

this condition being similar to that in Art. (**102**), we infer by the same reasoning that

$$\pm \frac{dz}{dx} \pm \frac{dz}{dy} = 0,$$

which cannot be for both the signs $\pm$ unless

$$\frac{dz}{dx} = 0 \text{ and } \frac{dz}{dy} = 0;$$

it being observed that owing to the independency of the variables, and the distinctive terms $\frac{dz}{dx}$ and $\frac{dz}{dy}$ depending severally on those variables, we are not entitled to election of signs like

$$+\frac{dz}{dx} - \frac{dz}{dy} = 0;$$

indeed, one of these terms need not exist (Art. **150**, **153**), but the demonstration must hold. Whatever term contains dy, must have contained k as factor.

Expunging, then, from the last inequation the terms $= 0$, the condition of z a maximum is,

$$\tfrac{1}{2}\left(\frac{d^2 z}{dx^2} \pm 2\frac{d^2 z}{dx\,dy} + \frac{d^2 z}{dy^2}\right) h^2 +, \text{etc.}, < 0;$$

and in the case of z a minimum, we should have derived in the same way,

$$\tfrac{1}{2}\left(\frac{d^2 z}{dx^2} \pm 2\frac{d^2 z}{dx\,dy} + \frac{d^2 z}{dy^2}\right) h^2 +, \text{etc.}, > 0.$$

Now x and y, and therefore z, and all the coefficients in fulfilling the above conditions, have determinate values; it is, therefore, determinable which of the last two inequations prevails in any given case, if we can avoid the complexity of the ambiguous sign $\pm$. Let us represent the terms within the parenthesis by

$$A \pm 2 B + C,$$

or

$$A \left(1 \pm 2 \frac{B}{A} + \frac{C}{A}\right);$$

adding $0 = \frac{B^2}{A^2} - \frac{B^2}{A^2}$ to the terms between the brackets, the expression becomes

$$A \left(\left(1 \pm \frac{B}{A}\right)^2 + \frac{C}{A} - \frac{B^2}{A^2}\right),$$

where the binomial is certainly plus, so that (since A and C agree in sign) if

$$\frac{C}{A} > \frac{B^2}{A^2}, \text{ that is, if } A\,C > B^2,$$

that is, if

$$A\,C - B^2 > 0,$$

the whole expression within the last double brackets will agree with A in sign. Hence, if

$$\left(\frac{d^2 z}{d x^2} \times \frac{d^2 z}{d y^2} - \left(\frac{d^2 z}{d x\, d y}\right)^2\right) > 0,$$

there is certainly a maximum or minimum, the former if $\frac{d^2 z}{d x^2} < 0$, the latter if $\frac{d^2 z}{d x^2} > 0$.

156. A function of two independent variables, x and y, may have a maximum as to one variable, and a minimum as to the other, at the same time; or a maximum or minimum as to one variable, and neither as to the other.

157. In functions of *two* independent variables, there are evidently eight conditions of value in regard to possible related positive and negative signs of the function and each respective variable (see Art. **97**), that is, as many conditions as the associate quantities $\pm z, \pm x, \pm y$ can be written in different ways, with the single sign *plus* or *minus* to each severally and independently. When either z, x, or y is at zero, such is a value of transition from one sign to the other; neither sign being significant when placed before zero.

SECTION XVIII.

PROBLEMS INVOLVING FUNCTIONS OF TWO INDEPENDENT VARIABLES; AND CASES OF THEIR MAXIMA AND MINIMA.

1. A person appropriated one day 10 dollars in payment for provisions, that were to be distributed in equal portions to some needy families; the next day he benefited, by gifts of clothing, 27 times as many families as each of those families of the previous day received pounds of provisions; on the third day he benefited, by fuel, 8 times as many families as those provisions had cost cents per pound. Required the smallest number of families, in all, which by any possibility may have received his aid on the three days.

Let $x =$ number of pounds to a family,

and $y =$ value in cents per pound of the provisions,

and $z =$ the whole number of families required;

$$\therefore \frac{1000}{xy} + 27x + 8y = z;$$

$$\therefore -\frac{1000}{x^2 y} + 27 = \frac{d z}{d x};$$

$$\therefore -\frac{1000}{x y^2} + 8 = \frac{d z}{d y}.$$

Now in case $z =$ max. or min., we have $\frac{d z}{d x} = 0$ and $\frac{d z}{d y} = 0$, from which we derive $x = 3.31$, $y = 11.17$, and $z = 215.71$. Now to determine whether this is a maximum or minimum value for z, we have

$$\frac{2000}{x^3 y} = \frac{d^2 z}{d x^2},$$

$$\frac{2000}{x y^3} = \frac{d^2 z}{d y^2},$$

and $$+\frac{1000}{x^2 y^2} = \frac{d^2 z}{d x\, d y};$$

so that for $$\frac{d^2 z}{d x^2} \times \frac{d^2 y}{d y^2} - \left(\frac{d^2 z}{d x\, d y}\right)^2 > 0,$$

we have $$\frac{2000}{x^3 y} \times \frac{2000}{x y^3} - \left(\frac{1000}{x^2 y^2}\right)^2 > 0;$$

also $$\frac{d^2 z}{d x^2} > 0, \text{ that is, } \frac{2000}{x^3 y} > 0;$$

hence the value 215.71 for z is a minimum.

Rational considerations will evidently, in the above case, enable us to determine whether it be a maximum or a minimum for z, if there be but one of them; for in the case of

$$\frac{1000}{x y} + 27\, x + 8\, y = z,$$

we see at once there is no limit to the greatness of z when either x or y becomes excessively great; and $\frac{1000}{x y}$ in positive values of x and y is never negative, however great x or y may be, and consequently however small the term may thus be rendered.

2. It is required to divide the number 48 into three such parts that the continued product of the first, the second power of the second, and third power of the third, may be the greatest possible. Ans. 8, 16, and 24.

3. A contractor agrees to fence the four sides of a rectangular field, but with two kinds of fence, one worth 78 cents per rod, the other worth \$1.25 per rod, and opposite sides to have a like fence; and he agrees to dig out rocks from one square rod of the field, this work being worth at the rate of \$7912 for the whole field. Required the length and width of the field, the cost of digging out the rocks, the cost of the whole fence on the sides and on the ends; when the sum of money that pays for the whole is the smallest, and required that sum.

Ans. In part, width of lot 12.585 rods.

4. A manufacturer of tin ware agrees to construct a tin box of rectangular sides and bottom, and without a top, and to hold just $5\frac{1}{2}$ cubic feet, with the least sheet tin possible. Required the dimensions and surface.

Let the bottom be x by y, then the height $= \dfrac{5\frac{1}{2}}{x\,y}$; then if $z =$ the surface, we have

$$z = x\,y + \frac{11}{y} + \frac{11}{x}.$$

Ans. x must $= y = (11)^{\frac{1}{3}} = 2.224$ feet, and height $= \frac{1}{2}\,(11)^{\frac{1}{3}} = 1.112$ feet, $\therefore$ the box is one half of a cube cut parallel to the bottom. But the whole cube might be cut any how by a plane through its centre, without a variation of the amount of surface or of the contents.

5. A miner in California dug uniformly some ounces of gold per day, for some days, when, becoming one of a company, consisting in all of as many miners as he had worked days alone, he received his share of 500 ounces, when the entire company changing to as many miners as he had dug ounces per day alone, he received his share of $342\frac{682}{729}$ ounces. After giving just this information to a speculator, the latter agreed to pay him for 185 ounces of gold for his all. Construing these conditions most favorably to the speculator, can he gain any thing?

Ans. He must lose the value of $18\frac{1}{3}$ ounces.

6. Some farmers bartered animals: in exchange for 5 heifers, Smith gave Jones 7 sheep and 3 dollars; Johnson gave Taylor 4 heifers and 2 dollars for 6 sheep, and Simpson gave Thomson a heifer for a sheep. After these transactions an army agent purchased all these same animals as of an approved and standard value, each kind; which gave rise to conversation among them as to who gained in their mutual trades. The three who gained, each agree to multiply the number of dollars gained, by itself, add the products together as so many dollars, and give this sum of dollars to the person who would tell them what it would be when it was the smallest it could be for any value of those animals, as that standard value; and required that sum.

If $x =$ number dollars value of a sheep,

and $y =$ number dollars value of a heifer,

and $z =$ that sum of money required;

then $z = (7x - 5y + 3)^2 + (6x - 4y - 2)^2 + (x - y)^2$,

or, all the signs of the quantities within the parentheses may be changed, since it will not affect z, and since we cannot presume that x is either greater or less than y.

Ans. A heifer, $5\frac{1}{6}$ dollars; a sheep, $3\frac{1}{2}$; Smith, Johnson, and Simpson gained $1\frac{2}{3}$ dollars each; sum required $8\frac{1}{3}$ dollars.

The following problems, (*a.*) and (*b.*), contain each but one independent variable:

7. (*a.*) A carriage wheel, which was in circumference 5 times the length of step of a certain pedestrian, and 1 foot more, the length of that step being $2\frac{1}{2}$ feet, ran once over a route 10 times as long as that between Dock Square in Boston and a certain station A, and then ran 1000 feet more. A second wheel, of such size that it would revolve 500 times in going once between Dock Square and station A, ran once between Bowdoin Square and station B, a distance equal to 1200 of those steps. Required the distance from Dock Square to A, when z, the sum of all the revolutions of the two wheels, is a minimum or maximum.

Ans. Distance 1679 feet, and z is a minimum, it being then 2501.4.

(*b.*) A carriage wheel, which was in circumference 5 times the length of step of a certain pedestrian, and 1 foot more, ran once over a route 10 times as long as that between Dock Square and a certain station A (which was a distance of 1679 feet), and then ran 1000 feet more. A second wheel, of such size that it would revolve 500 times in going once between Dock Square and station A, ran once between Bowdoin Square and station B, a distance equal to 1200 of those steps. Required the length of that step when the sum z, of all the revolutions of the two wheels, is a minimum.

Ans. Step 3.28 feet; now z is a minimum, at 2500.8.

The following problem (*c.*) is the same as (*a.*) and (*b.*) preceding, except that it combines in one (function of x and y) $= z$, each of the same variables as in (*a.*) and (*b.*); thus making two independent variables:

(*c.*) A carriage wheel, which was in circumference 5 times the length of step of a certain pedestrian, and 1 foot more, ran once over a route 10 times as long as that between Dock Square and a certain station A, and then ran 1000 feet more. A second wheel, of such size that it would revolve 500 times in going once between Dock Square and station A, ran once between Bowdoin Square and station B, a distance equal to 1200 of those steps. Required the distance from Dock Square to A, and the length of that step when z, the sum of all the revolutions of the two wheels, is a maximum or minimum.

8. (*a.*) A certain perpendicular flag-staff, 129 feet high, stands on a level plain; a stake is driven into the ground to mark a point 60 feet to the south of the base of that staff, which point is joined with the top of the staff by a straight cord. Another such flag-staff, 97 feet high, stands 82 feet to the westward of the first, and its top is joined by a cord to a point at the surface of the ground marked by a stake 40 feet to the east of the first stake. Required to determine the nearest distance between one cord and the other, either produced indefinitely if necessary, which might be the case in a generalization of the conditions.

It will be useful, in the solution of the above problem, to conceive three arbitrary planes cutting each other at right angles, to each of which any point in either cord may be referred by perpendicular measurement; through the medium of right-angled triangles, an expression may be found for a perpendicular line from one cord to the other; this line must be a minimum. By means of a solid diagram, constructed of pasteboards for planes and threads for lines, there is much simplicity in the solution.

(*b*.) Required to determine the diameter of the smallest sphere to which the above two lines are tangent, and whether this diameter is the line required in the foregoing question.

The following problems trespass upon the rule hitherto adhered to, not to propose geometrical and trigonometrical problems, except the most elementary.

9. A sentinel has received orders from his commanding officer to visit in succession three important posts, A, B, and C, and return from each visit of each to his camp. The post A is 90 rods from B, B 23 rods from C, and C 72 rods from A. But he may place his camp where he pleases. Required the distance of his camp severally from A, B, and C, when a round of visits is made with the fewest steps, and consequently any number of rounds, the ground supposed level.

The following problem is to be considered general, with reference to Sections XII., XVI., and XVIII.

10. A plain is level, and the sight over it is unobstructed by objects: on it is a circular course 80 rods in diameter. A procession was once seen marching round this course, in which was a person carrying a banner 4 feet square, and holding it perpendicularly, with its centre 9 feet above the plain, and 40 rods, horizontally measured, from the centre of the course. The banner was constantly held with its plane agreeing with the radius of this course, and hence invisible to a person at the centre of the course. But outside there was a stationary observer, so situated that his eye was 70 rods from the centre of the course, and 9 feet above the plain. As the banner was thus carried entirely

round this course, its two sides being successively exhibited to that observer, it is probable that there were two locations, one toward the right, the other toward the left of the observer, where that banner appeared the largest object, as when it should be projected, when any where on its circuit, to form a portion of the surface of a sphere, of which the observer's eye is at the centre, such surface being assumed at any distance whatever. Required the distance in rods from the eye to centre of banner, when the banner appeared largest.

The planet Venus gives her maximum light to the earth on conditions not much unlike the above.

SECTION XIX.

DEMONSTRATION, OF THE GENERAL FORM OF THE DEVELOPMENT OF $f(x+h)$, AND, OF THE DIFFERENTIATION OF CERTAIN FUNCTIONS.

158. We have deferred to the present section a very elementary and important demonstration in regard to the form of the development of any function of x whatever, when x becomes $x+h$, and our earlier endeavors to illustrate the nature and rules of differentiation were at a disadvantage on account of the omission.

Let fx be any function of x, and when x becomes $x+h$, $f(x+h)$ will have a general development of the form

$$f(x+h)=fx+Ah+Bh^2+Ch^3+\text{, etc.,}$$

in which A, B, C, etc., are coefficients containing x and constants, and each may evidently be an aggregate of certain sub-terms, etc. The form is intended to show how h appears in the development.

It is useful to prove the foregoing general development without reference to the binomial or any other theorem.

One of the terms of the general development must be $f\,x$, in which h in no respect exists as a factor or otherwise, because when $h = 0$, and $f\,(x + h)$ becomes $f\,x$, the development ought to reduce to

$$f\,x = f\,x.$$

Nor can there be any term but $f\,x$ in the development which does not contain h as a factor.

None of the indexes of h can be negative; because if h have a negative index, h may be made to appear as a denominator of A, B, or C, etc., with that index positive. Such term would, therefore, become infinite when $h = 0$, but when $h = 0$ the term itself ought to become 0, because the condition of the equation becomes $f\,x = f\,x$, and $f\,x$ is not necessarily infinite, nor has it any restricted value, nor any value, therefore, that ought to be restricted in the development. Nor can there be in the series two terms, each infinite and with opposite signs, because they would not be *equal infinites* unless they should be rendered so by the vanishing of h at like powers or like rates, and in such case such two terms become one in the series.

None of the exponents of h can be fractional, because of one factor of such term, a root is indicated to be taken. Now all these terms of the *general* development are supposed to be numerical; and the roots only of *particular* numerical quantities are rational, such as 1, $\frac{1}{4}$, $\frac{1}{9}$ 4, 16, 8, 27, etc.; the roots of intermediate quantities may be irrational or inexpressible in number. Therefore the roots of numerical quantities *in general* are irrational. This truth is not affected by the consideration that the *roots of powers* are indicated by fractional exponents. Now if a root of h be irrational, the term which contains it is irrational. Notwithstanding what the character of the other terms of the development may be, $f(x+h)$ being equated with an expression, one of the terms of which is irrational, is itself irrational in *general values;* hence a restriction is imposed upon the values of the development of $f(x+h)$, and upon fx, and such development is not general.

The indexes of h in the successive terms, are the natural series of entire and positive numbers, 1, 2, 3, etc.; for if A be the coefficient of h at the lowest power or a, we may write the development thus:

$$f(x+h) = fx + (A + B h^{b-a} + C h^{c-a} +, \text{etc.,})\, h^a;$$

but which for simplicity we will write thus:

$$f(x+h) = fx + (A+P)\, h^a;$$

$$\therefore \frac{f(x+h) - fx}{A+P} = h^a;$$

$$\therefore \frac{(f(x+h) - fx)^{\frac{1}{a}}}{(A+P)^{\frac{1}{a}}} = h;$$

wherefore if a is not a unit, we have $f(x+h) - fx$ rendered irrational, and h itself irrational, and consequently each at restricted values, which are opposed to the hypothesis; therefore a is unity.

Since a is unity, we have

$$f(x+h) - fx = (A+P)\, h^1;$$

or, rather,

$$f(x+h) - fx - A h^1 = B h^{b-1} + C h^{c-1} +, \text{etc.}$$

Again, as before,

$$f(x+h) - fx - A h^1 = (B + C h^{c-b-1} +, \text{etc.}), h^{b-1};$$

but which, for simplicity, we will put

$$f(x+h) - fx - A h^1 = (B + Q) h^{b-1};$$

$$\therefore \frac{(f(x+h) - fx - A h^1)^{\frac{1}{b-1}}}{(B+Q)^{\frac{1}{b-1}}} = h;$$

which imposes restrictions against hypothesis unless $b=2$, for when $b=2$, we find $\frac{1}{b-1}=1$. Now, by the continuation of this course of reasoning, we may show $c=3$, etc. *Therefore the indexes*, etc.

If h be negative as in $f(x-h)$, it is evident that the general development is the same in form, except that the terms having h with indexes odd, will be negative.

The coefficients A, B, C, etc., in the general development, are evidently functions of x, but are not as yet, except A, differential coefficients. (Art. **99.**)

This development being general, holds for such particular cases as render one or more of its terms imaginary.

Although authors are very reserved, and some of them entirely silent, respecting the possible value of h in this development, it is plain that it must be an infinitesimal, or indefinitely small. If in the formula, $2h$ be substituted for h, it becomes

$$f(x+2h) = fx + 2Ah + 4Bh^2 + 8Ch^3 +, \text{etc.},$$

which shows that the coefficients A, B, C, etc., are affected by, and rendered dependent upon, a change of the value of h.

159. By transposing the first term in the general development of $f(x+h)$, we have

$$f(x+h) - fx = Ah + Bh^2 + Ch^3 +, \text{etc.}$$

Now the first member of this equation is the amount that

$f\,x$ changes by virtue of h added to the variable x. Dividing by h, the increment of x, we have the fractional or proper form of expressing the ratio of the change of value of the function to that of the variable, that is,

$$\frac{f(x+h)-f\,x}{h} = A + B\,h + C\,h^2 +, \text{ etc.}$$

Now when $h = 0$, the numerator may take the designation $d\,f\,x$ (or if $f\,x = y$), of $d\,y$, and the denominator h must take the designation $d\,x$.

$$\therefore \frac{d\,y}{d\,x} = A,$$

which we may enunciate thus: *the coefficient of the second term of the general development of $f\,(x+h)$ is the dif. coef. derived from $f\,x$.* (Art. 68.)

160. The form of the general development of $f\,(x+h)$ furnishes the means of a formal demonstration of *the method of differentiating the product of two or more functions of the same variable.*

Let y and z be functions of x in the expression

$$u = a\,y\,z.$$

By changing x into $x + h$, the function y becomes (designating by y' the new value of y)

$$y' = y + A\,h + B\,h^2 + C\,h^3 +, \text{ etc.,} \qquad (1.).$$

and the function z becomes

$$z' = z + A'\,h + B'\,h^2 + C'\,h^3 +, \text{ etc.} \qquad (2.).$$

Hence, when $h = 0$, we have from (1.),

$$\frac{y'-y}{h} = \frac{d\,y}{d\,x} = A;$$

and from (2.),

$$\frac{z' - z}{h} = \frac{d z}{d x} = A'.$$

Designating by u' the new value of u received in consequence of the change of y and z, and multiplying the product of (1.) and (2.) by a, we have

$$u' = a y z + a (A z + A' y) h +, \text{etc.},$$

$$= a y z + a \left(\frac{d y}{d x} z + \frac{d z}{d x} y\right) h +, \text{etc.};$$

therefore $a \left(\frac{d y}{d x} z + \frac{d z}{d x} y\right)$ being the coef. of the second term of the development of u', we have

$$\frac{d u}{d x} = a z \frac{d y}{d x} + a y \frac{d z}{d x};$$

$$\therefore d u = a z d y + a y d z. \qquad (3.).$$

Hence, *to differentiate the product of two functions of the same variable, we must multiply each by the differential of the other, and add the results.*

161. It will be easy now to express the differential of a product of three functions of the same variable. Let

$$u = w y z$$

be a product of three functions of x; then, putting v for $w y$, the expression is

$$u = v z;$$

hence by (3.),

$$d u = z d v + v d z,$$

but $v = w y$; therefore by (3.),

$$d v = y d w + w d y;$$

consequently by substitution

$$d u = z y d w + z w d y + w y d z;$$

and it is plain that in this way the differential may be found, be the factors ever so many; so that generally, *to differentiate a product of several functions of the same variable, we must multiply the differential of each factor by the product of all the other factors, and add the results.* (Arts. 74, 75.)

162. If it be required to differentiate an expression consisting of several functions of the same variable combined by addition or subtraction, it will be necessary merely to differentiate each separately, and to connect together the results by their respective signs. For let the expression be

$$u = a w + b y + c z +, \text{ etc.},$$

in which w, y, and z are functions of x. Then, changing x into $x + h$, and developing,

w becomes $w + A h + B h^2 +$, etc.,

y " $y + A' h + B' h^2 +$, etc.,

z " $z + A'' h + B'' h^2 +$, etc.,

$\therefore u$ " $u + (a A + b A' + c A'' +, \text{etc.}), h +$, etc.,

$$\therefore d u = a A d x + b A' d x + c A'' d x +, \text{ etc.}$$

But $A d x = d w$, $A' d x = d y$, $A'' d x = d z$, etc.;

$$\therefore d u = a d w + b d y + c d z +, \text{ etc.};$$

that is, the differential of the sum of any number of functions is equal to the sum of their respective differentials. (Art. 76.)

SECTION XX.

MACLAURIN'S THEOREM, AND ITS APPLICATION.

163. This theorem gives a general mode of developing, expanding, or changing the form of, some algebraic (and other) expressions by a series arranged with reference to the positive ascending powers of any one specific quantity in them, which may be assumed for the purpose.

A function having a single variable is such an expression, and the variable may be selected as the specific quantity in question. Since only the form is changed, the value of the expression, or of the function (if it have a value) must remain unchanged, except in marked exceptional cases. If a function has no specific value, the new form of it produced by this theorem must have the same range of values, if the values are real, as the original function. Sometimes the odd, sometimes the even powers of the specific quantity become eliminated from the series, because the coefficients of such terms must respectively equal zero. If x be the quantity, or represent the position of the quantity according to the ascending powers of which the series is to be formed, then the expression being called a function of x, and A, B, C, etc., being indeterminate coefficients, successively, of the powers of x, we have

$$F x = y = A + B x + C x^2 + D x^3 + E x^4 +, \text{ etc.};$$

$$\therefore \frac{d y}{d x} = B + 2 C x + 3 D x^2 + 4 E x^3 +, \text{ etc.};$$

$$\therefore \frac{d^2 y}{d x^2} = 2 C + 2 . 3 D x + 3 . 4 E x^2 +, \text{ etc.};$$

$$\therefore \frac{d^3 y}{d x^3} = + 2 . 3 D + 2 . 3 . 4 E x +, \text{ etc.};$$

$$\therefore \frac{d^4 y}{d x^4} = 2.3.4\, E, \text{ etc.}$$

Now, if, by making $x = 0$, we select the particular values respectively for these coefficients, and the function, *which are not affected by the value of* x, and place them in parentheses to denote this, we have

$$(y) = A;$$

$$\left(\frac{d y}{d x}\right) = B;$$

$$\left(\frac{d^2 y}{d x^2}\right) = 2\, C \therefore C = \left(\frac{d^2 y}{d x^2}\right) \frac{1}{2};$$

$$\left(\frac{d^3 y}{d x^3}\right) = 2.3\, D \therefore D = \left(\frac{d^3 y}{d x^3}\right) \frac{1}{1.2.3};$$

$$\left(\frac{d^4 y}{d x^4}\right) = 2.3.4\, E \therefore E = \left(\frac{d^4 y}{d x^4}\right) \frac{1}{1.2.3.4}.$$

Substituting these expressions for A, B, C, etc., the series becomes

$$F x = (y) + \left(\frac{d y}{d x}\right) x + \left(\frac{d^2 y}{d x^2}\right) \frac{x^2}{1.2} + \left(\frac{d^3 y}{d x^3}\right) \frac{x^3}{1.2.3} +, \text{ etc.,}$$

which is Maclaurin's Theorem.

Although, to derive these coefficients, x was made equal to zero, yet by hypothesis they are such as the value of x cannot affect; therefore, in the theorem, x may be restored to any value consistent with the function.

Otherwise:

Taylor's Theorem being,

$$F(x + h) =$$

$$F x + \frac{d F x}{d x} \frac{h}{1} + \frac{d^2 F x}{d x^2} \frac{h^2}{1.2} + \frac{d^3 F x}{d x^3} \frac{h^3}{1.2.3} +, \text{ etc.;}$$

if $x = 0$, it becomes

$$F h =$$

$$(F x) + \left(\frac{d F x}{d x}\right) \frac{h}{1} + \left(\frac{d^2 F x}{d x^2}\right) \frac{h^2}{1.2} + \left(\frac{d^3 F x}{d x^3}\right) \frac{h^3}{1.2.3},$$

where the parentheses are used to intimate that x has this restricted value, $x = 0$. These differential coefficients, in parentheses, are constant, for in the *actual* coefficients of a particular case applied under this general notation, x will not be found. Therefore, h is no longer limited to the value 0, but may be of any greatness; and since x has disappeared, we may revive it in h. We then have, calling the original $F x = y$ and $(F x) = (y)$,

$$F x =$$

$$(y) + \left(\frac{d y}{d x}\right) x + \left(\frac{d^2 y}{d x^2}\right) \frac{x^2}{1.2} + \left(\frac{d^3 y}{d x^3}\right) \frac{x^3}{1.2.3} +, \text{ etc.,}$$

which is Maclaurin's Theorem.

1. Required the deyelopment of $y = F x$, viz., $\frac{1}{a + x}$.

$$\frac{d y}{d x} = -\frac{1}{(a + x)^2} \therefore \left(\frac{d y}{d x}\right) = -\frac{1}{a^2};$$

$$\frac{d^2 y}{d x^2} = +\frac{2}{(a + x)^3} \therefore \left(\frac{d^3 y}{d x^3}\right) = \frac{2}{a^4};$$

$$\frac{d^3 y}{d x^3} = +\frac{2.3}{(a + x)^4} \therefore \left(\frac{d^3 y}{d x^3}\right) = -\frac{2.3}{a^4};$$

$$\therefore \frac{1}{a + x} = \frac{1}{a} - \frac{x}{a^2} + \frac{x^2}{a^3} - \frac{x^2}{a^4} +, \text{ etc.}$$

2. Required $y = (b^2 + x^2)^{\frac{1}{2}}$ in a series.

$$\text{Ans. } y = b + \frac{x^2}{2 b} - \frac{x^4}{2.4.b^3} + \frac{3 x^6}{2.4.6 b^5} -, \text{ etc.}$$

4. It is required to express in a series of the ascending positive powers of x this function of x, viz., $\frac{a}{b-x}$.

$$\text{Ans. } \frac{a}{b-x} = \frac{a}{b} - \frac{a}{b^2}x + \frac{a}{b^3}x^2 - \frac{a}{b^4}x^3 +, \text{ etc.}$$

5. It is required to develop the algebraic expression $\sqrt{(a^2+b^2)}$ in a series with reference to the increasing positive powers of b.

$$\text{Ans. } \sqrt{(a^2+b^2)} = a + \frac{b^2}{2a} - \frac{b^4}{8a^3} + \frac{b^6}{16a^5} -, \text{ etc.}$$

6. Change the form of $\frac{c}{b-ax}$ as a function of x.

$$\text{Ans. } \frac{c}{b-ax} = \frac{c}{b}\left(1 + \frac{ax}{b} + \frac{a^2x^2}{b^2} + \frac{a^3x^3}{b^3} +, \text{etc.}\right)$$

7. Develop, if possible, $y = ax^4$ by Maclaurin's Theorem.

$$y = ax^4, \therefore (y) = 0;$$

$$\frac{dy}{dx} = 4ax^3, \therefore \left(\frac{dy}{dx}\right) = 0;$$

$$\frac{d^2y}{dx^2} = 12ax^2, \therefore \left(\frac{d^2y}{dx^2}\right) = 0;$$

$$\frac{d^3y}{dx} = 24ax, \therefore \left(\frac{d^3y}{dx^3}\right) = 0;$$

$$\frac{d^4y}{dx^4} = 24a, \quad \therefore \left(\frac{d^4y}{dx^4}\right) = 24a;$$

$$\therefore ax^4 = 0 + 0 + 0 + 0 + 24a\frac{x^4}{1.2.3.4} + 0, \text{ etc.}$$

Whence it appears that in this case the theorem does not fail in truthfulness, but in utility. The development is not a series.

8. Develop into a series $y = \frac{1}{1-x}$.

Ans. $y = 1 + x + x^2 + x^3 + x^4 +$, etc.

Whence it appears that the theorem fails to give an equivalent substitute for y when $1 < x$, because the substitute becomes infinite.

It may have been noticed that all the results obtained in the foregoing examples might also be obtained by ordinary algebraic division or extraction of roots, or at least by the Binomial Theorem.

But we ordinarily do not have an algebraic demonstration of this theorem so general as to embrace developments in case the indexes are fractional, denoting both a power and a root, as $\frac{2}{3}$. Although this theorem embraces the binomial, it is, therefore, more general. It is the foundation of other theorems like Lagrange's, and appears indispensable in the higher Calculus, and in trigonometrical analysis.

9. Let it be required to develop

$$y = \frac{x+1}{a + (c^2 + x^2)^{\frac{1}{2}}}.$$

10. Let it be required to develop

$$y = \frac{x^3 - 1}{a x + \sqrt{x}}.$$

164. *Binomial Theorem.* If the expression $(a+x)^n$ be developed by Maclaurin's Theorem, the result exhibits the Binomial Theorem.

11. Let it be required to develop Fx, or $y = (a + x)^n$, according to the ascending positive integral powers of x, or by Maclaurin's Theorem:

We have $\quad y = (a + x)^n, \ \therefore (y) = a^n,$

$$\frac{dy}{dx} = n\,(a + x)^{n-1},$$

$$\therefore \left(\frac{dy}{dx}\right) = n\,(a)^{n-1};$$

$$\frac{d^2 y}{dx^2} = n\,(n - 1)\,(a + x)^{n-2},$$

$$\therefore \left(\frac{d^2 x}{dy^2}\right) = n\,(n - 1)\,a^{n-2};$$

$$\frac{d^3 y}{dx^3} = n\,(n - 1)\,(n - 2)\,(a + x)^{n-3},$$

$$\therefore \left(\frac{d^3 y}{dx^3}\right) = n\,(n - 1)\,(n - 2)\,a^{n-3};$$

$$\therefore \frac{d^n y}{dx^n} = n\,(n - 1)\,(n - 2)\,(n - 3)\ldots(n - n)\,(a + x)^{n-n}$$

Now, y being Fx with reference to Maclaurin's Theorem, we have $\left(\frac{dy}{dx}\right) = \left(\frac{dFx}{dx}\right); \left(\frac{d^2 y}{dx^2}\right) = \left(\frac{d^2 Fx}{dx^2}\right)$, etc.; hence, substituting the equivalents of these as found, we have

$$Fx = y = (a + x)^n = a^n + n a^{n-1} x + \frac{n\,(n - 1)}{2} a^{n-2} x^2 + \frac{n\,(n - 1)\,(n - 2)}{2\,.\,3} a^{n-3} x^3 +, \text{etc.},$$

which is the *Binomial Theorem*, and n may be a whole number or a fraction, or be negative or irrational. It will be observed that when n is a whole number, since $n - n$

becomes a factor in the nth dif. coef., it renders it $= 0$, anc the series must terminate.

In case the index n should not be a whole number, bu should be a fraction like $\frac{p}{q}$ when in its simplest form, thei the differences $\frac{p}{q}-1$, $\frac{p}{q}-2, \ldots \frac{p}{q}-n$, etc., by the suc cessive subtractions of the integral numbers $1, 2, 3 \ldots n$ could never be reduced to 0, but would pass over 0 i going from $+$ to $-$ values; and the series would neve terminate.

165. But the mode of developing, by Maclaurin's Theo rem, a function of x, is applicable as well when the func tion is implicit. It may be observed that, when x i made $= 0$, y being dependent will take some value i constants corresponding to $x = 0$, which value is to b substituted for y.

12. Let it be required to develop y according to th ascending powers of x in

$$y^3 - 3y + x = 0;$$

$$3y^2\,dy - 3\,dy + dx = 0;$$

$$\frac{dy}{dx} = \frac{1}{3y^2 - 3}, \quad \therefore \left(\frac{dy}{dx}\right) = \frac{1}{3};$$

$$\frac{d^2y}{dx^2} = \frac{6y\frac{dy}{dx}}{(3y^2-3)^2}, \quad \therefore \left(\frac{d^2y}{dx^2}\right) = 0;$$

$$\frac{d^3y}{dx^3} = \frac{6(3y^2-3)^2\left(\frac{dy}{dx}\cdot\frac{dy}{dx} + y\frac{d^2y}{dx^2}\right) - 6y\,d(3y^2-3)^2\,\frac{dy}{dx^2}}{(3y^2-3)^4};$$

$$\therefore \left(\frac{d^3y}{dx^3}\right) = \frac{6 \times 3^2(\frac{1}{3} \times \frac{1}{3})}{3^4} = \frac{6}{3^4};$$

$$\therefore y = \frac{x}{3} + \frac{x^3}{3^4} + \frac{x^5}{3^6} +, \text{ etc.}$$

It may be observed that the *differential* of the numerator $6y\frac{dy}{dx}$, this being a product of the two variables y and dy, dx as well as $\frac{1}{dx}$, being constant, and d^2y being the differential of dy, is

$$6y\frac{d^2y}{dx}+6\frac{dy}{dx}\times dy,$$

which terms, as well as the others of the new numerator which is being formed, are to be divided by dx, as well as the other member of the equation, which is now $\frac{d^3y}{dx^2}$; hence we have, in the third dif. coef., as a factor

$$6\left(\frac{dy}{dx}\cdot\frac{dy}{dx}+y\frac{d^2y}{dx^2}\right).$$

Since $y=0$ when $x=0$, all terms containing y as a factor become eliminated in finding those dif. coefs. within brackets.

13. Let it be required to develop y in $by^3-xy=b$, according to the ascending powers of x.

$$\text{Ans. } y=1+\frac{x}{3b}-\frac{x^3}{3^4b^3}+a, \text{ etc.}$$

14. Let it be required to develop y in $y^3x-a^3(y+x)=0$, according to the powers of x.

$$\text{Ans. } y=-x-\frac{x^4}{a^3}-\frac{3x^7}{a^6}-\text{etc.}$$

166. It should not escape notice that here is shown an achievement, by the analysis of the calculus, beyond the ordinary direct resources of algebra, in relation to the resolution of equations, and is particularly available whenever the series generated in the manner shown, is so converging that the terms of it may be readily summed without tediousness, as when the denominators, when there are any, are large relatively to the numerators, or to x, which may always be in a numerator, when the terms are fractional.

Nor should it escape notice what may be accomplished by the transformation of x into y, or y into x, or of any known constant into x or y, or by creating x at pleasure, with the substitution of it for any term or quantity, for the sake of the development; and of the means of equating any unknown quantity situated as y, with others which are presumed to be known.

It is useful to observe that these general analyses never fail to embrace the *truth* of quite elementary conditions, even when a *series* may fail to have place. For,—

15. Let it be required to develop y in $y^3 - x^3 = 0$, according to the powers of x.

We have
$$y = x, \therefore (y) = 0;$$
$$\frac{dy}{dx} = \frac{3x^2}{3y^2} \therefore \left(\frac{dy}{dx}\right) = 1;$$
$$\frac{d^2y}{dx} = \frac{6x\,dx \times 3y^2 \quad - 3x^2 \quad \times 6y\,dy}{9y^4},$$
$$\frac{d^2y}{dx^2} = \frac{6x \times 3y^2 \quad - 3x^2 \times 6y\frac{dy}{dx}}{9y^4},$$
$$\therefore \left(\frac{d^2y}{dx^2}\right) = \frac{0}{0} = 0,$$
$$\therefore y = 0 + x + 0 +, \text{etc.};$$

where we are obliged to remark that the foregoing $\frac{3x^2}{3y^2}$, as well as the second dif. coef., in its general form, are reduced, respectively, to 1 and 0, from expressions each virtually $\frac{0}{0}$ when $x = 0$, by principles that will be fully demonstrated in the following section. At present it is sufficient to say that

$$\frac{3x^2}{3y^2} = 1,$$

because both numerator and denominator become *alike* when $x = 0$, and the second dif. coef., that is,

$$6x \times 3y^2 - 3x^2 \times 6y = 0$$

when x and y are *any how* alike, without notice of the circumstance that their respective values are 0, which *casually* renders $9y^4 = 0$.

167. Maclaurin's Theorem fails to give a true development of all functious of x, of which any dif. coef. becomes infinite when $x = 0$. We know nothing of a development by it, when x should be restricted to the value 0, and of which any dif. coef. becomes infinite. And since in the theorem, x in the position of all its ascending powers is not, by the nature of the theorem, to be restricted to any value, when $0 < x$ any such term, and, as will be seen by a few examples, all succeeding terms become infinite. Such, therefore, cannot be a development of a function which is not necessarily infinite. Thus, if

$$y = x^{\frac{1}{2}},$$

$$\frac{dy}{dx} = \frac{1}{2x^{\frac{1}{2}}} = \infty \text{ when } x = 0,$$

$$\frac{d^2y}{dx^2} = -\frac{1}{4x^{\frac{3}{2}}} = -\infty \text{ when } x = 0.$$

So, also, with $a x^{\frac{1}{3}}$, $(ax - x^2)^{\frac{1}{2}}$, $b x^{\frac{5}{3}}$, etc. But the infinite dif. coef. might be deferred to the 5th, 6th, or the nth, for the obvious reason that several successive subtractions of unity from an improper fraction may be necessary before the remainder becomes negative.

168. In Arts. **151, 152**, we have given the development of a function of two independent variables by Taylor's Theorem. In the same manner in which we have derived Maclaurin's Theorem from Taylor's as for one variable, we may derive the development of a function of two independent variables by Maclaurin's. If in that development (Art. **152**) we suppose x and y each $= 0$, the development will become that of $F(h, k)$ according to the powers of h and k, or substituting in that development x for h and y for k, since these may now have any value, we have

$$z = F(x, y) = (z) + \left(\left(\frac{dz}{dx}\right) x + \left(\frac{dz}{dy}\right) y\right) +$$

$$\frac{1}{2}\left(\left(\frac{d^2 z}{dx^2}\right) x^2 + 2\left(\frac{d^2 z}{dx\,dy}\right) x y + \left(\frac{d^2 z}{dy^2}\right) y^2\right) + \text{etc.}$$

16. Let it be required to develop z according to the powers of x and y in

$$z = a x^2 (b - y^3) - y^2 (x^2 + c)^2.$$

SECTION XXI.

DETERMINATION OF THE VALUE OF VANISHING FRACTIONS.

169. We have, on several occasions, compared the rates of change of value of two functions of the same variable, for some particular value of the variable. In such case, the variable may not only be expressed as x in each function, but by hypothesis is to be the same x, and therefore is to have a common value in each function.

When such two functions become, respectively, the numerator and denominator of a fraction, the case may happen when, on the variable taking a particular value, the fraction reduces to the form $\frac{0}{0}$; such is called a *vanishing fraction.* This value is indeterminate in the abstract, but determinate when we know its origin. The *value* of a vanishing fraction does not necessarily vanish. The numerator and denominator vanish severally and independently, or by independent rates of change.

1. On an occasion it cost a man 75 cents per mile to travel; however, of the whole number of miles travelled, 42 were without cost. If a sum like that expended in this travelling, should be expended in the purchase of 25 times as many pounds of the commodity C as he had travelled miles with cost, what would have been its price per pound, whatever the number of miles travelled with cost might have been, even if it had been the least conceivable in a fraction?

Let $x =$ number of miles travelled in all; then the price per pound of C will be represented thus:

$$\frac{(x - 42)\,75}{(x - 42)\,25} = 3 \text{ cents.}$$

Here it is evident that, in case $x = 42$, the fraction reduces to $\frac{0}{0}$; but its value appears to be 3 nevertheless. If we watch the relative values of the numerator and denominator while, by a variation of x, they are becoming exceedingly small, it is quite evident that nothing disturbs the ratio of their values.

2. A courier travelled 15 or a hours, at 15 or a miles per hour, in a continuous course, when he travelled, in return, just as many hours as miles per hour; we need not say, as yet, whether or not he had accomplished just his return, but a messenger was ready to, or did proceed

to meet him at a rate per hour equal to the excess of the rate per hour of the courier's set-out, above the rate of his return, per hour. Required to determine the number of hours necessary for the messenger's travel, although the distance necessary for him to travel were the shortest conceivable.

Let $x =$ the miles per hour of the return; then $\frac{a^2 - x^2}{a - x} = \frac{(a + x)(a - x)}{a - x} = a + x = 30$ hours when $x =$ 15, and the number of hours required; and it appears that the messenger has farther to go, and goes in less time, the more a exceeds x. The numerator $a^2 - x^2$ being an expression of the second degree, does not vary uniformly with x, hence the quotient has a value that varies not uniformly, but approaches a fixed amount for $x = a$.

3. Required the value of $\frac{a\,x}{x}$ when $x = 0$.

Ans. $\frac{a\,x}{x} = \frac{a}{1} = a$.

170. In the cases which have thus far been presented we have evidently obtained the required value of the vanishing fractions by reducing the fraction to its lowest terms, or by simple algebraic processes.

4. Required the value of $\frac{\sqrt{x} - \sqrt{a}}{\sqrt{x^2 - a^2}}$ when $x = a$. In this case, if we divide the numerator and denominator by $\frac{1}{\sqrt{x} + \sqrt{a}}$ and the resulting quotients by $\sqrt{x - a}$, we have the following:

$$\frac{x - a}{\sqrt{(x^2 - a^2)}\,(\sqrt{x} + \sqrt{a})} = \frac{x - a}{\sqrt{x - a} \cdot \sqrt{x + a} \,.\, (\sqrt{x} + \sqrt{a})} =$$

$$\frac{\sqrt{x - a}}{\sqrt{x + a} \,.\, (\sqrt{x} + \sqrt{a})} = 0 \text{ when } x = a,$$

because we have removed all negative quantities from the

denominator; to produce the same result by one divisor, it must be

$$\frac{\sqrt{x-a}}{\sqrt{x}+\sqrt{a}};$$

a divisor that is very far from being quite obvious. So that it appears that these algebraic processes cannot have in general view the reduction of the fraction to its lowest terms, but to effect a transformation of whatever kind that may, as above, remove ambiguity. It is better, therefore, to adopt a direct and uniform process for determining the value of a vanishing fraction: this process the calculus, by differentiation, supplies.

171. Since all algebraic functions of a variable must vary in value when the variable does, for the variable is immediately eliminable from all expressions containing it, which do not vary when the variable does, such as $b+(a-a)\,x^3$, $\frac{33\frac{1}{3}\,x+2000}{x+60}$, etc., it follows that the successive differential coefficients of a function cannot all be 0 in value, or vanish for a particular value of the variable.

In the case of fractions vanishing at a particular value of the variable, we have evidently two functions of one and the same variable, and which need not be like in form, and which therefore ought to be designated by the distinctions of, say, Fx for the numerator, and fx for the denominator; this gives us, in view of the hypothesis,

$$\frac{Fx}{fx}=\frac{0}{0}.$$

Now, in case Fx and fx are of a nature to be developed by Taylor's Theorem for the value in question, let them be respectively developed, or let us entertain the suggestion of each of their values moving out, as it were, from zero by the nearest appreciable amount, as when x should be $x + h$; then we have, instead of zero for Fx, and for fx, certain indefinitely small compared quantities, if one or the other should not still remain zero; in which case, we have what we seek for, in 0 or ∞, as the value of the fraction. Understanding, now, that Fx is y in Taylor's Theorem, and that fx is some other and a different y, and agreeing, for convenience, to represent

$$\frac{dy}{dx} \text{ by } p', \ \frac{d^2y}{dx^2.1.2} \text{ by } p'', \ \frac{d^3y}{dx^3.1.2.3} \text{ by } p''', \text{ etc.}$$

$$\frac{dy}{dx} \text{ by } q', \ \frac{d^2y}{dx^2.1.2} \text{ by } q'', \ \frac{d^3y}{dx^3.1.2.3} \text{ by } q''', \text{ etc.,}$$

where the number of accents is made to agree with the *order* of the dif. coefs. in numerical name, we have

$$\frac{F(x+h)}{f(x+h)} = \frac{Fx + p'h + p''h^2 + p'''h^3 + \text{etc.}}{fx + q'h + q''h^2 + q'''h^3 + \text{etc.}}, \qquad (1.)$$

Now, since by hypothesis $Fx = 0, fx = 0$, at the value in question, they severally become of no account in the fraction of the development, and may be expunged, so that we have, after dividing by h,

$$\frac{F(x+h)}{f(x+h)} = \frac{p' + p''h + p'''h^2 + \text{etc.}}{q' + q''h + q'''h^2 + \text{etc.}}; \qquad (2.)$$

and when $h = 0$,

$$\frac{Fx}{fx} = \frac{p'}{q'},$$

as the required value of the vanishing fraction; but possibly $\frac{p'}{q'}$ may become $\frac{0}{0}$, in which case we may expunge $\frac{p'}{q'}$ from equation (2.), which then becomes, after dividing by h,

$$\frac{F(x+h)}{f(x+h)} = \frac{p'' + p'''h + \text{etc.}}{q'' + q'''h + \text{etc.}}, \qquad (3.)$$

which becomes, when $h = 0$,

$$\frac{Fx}{fx} = \frac{p''}{q''},$$

as the required value in case it does not become $\frac{0}{0}$; in which case expunge $\frac{p''}{q''}$ from (3.), and divide by h, and we have, when $h = 0$,

$$\frac{Fx}{fx} = \frac{p'''}{q'''},$$

as the required value, if it is any thing else than $\frac{0}{0}$; and so on, so that we have the following rule for determining the value of a fraction of which the numerator and denominator vanish when x takes a particular value:

172. *For the numerator and denominator substitute their first dif. coefs., their second dif. coefs., and so on, till we obtain the first fraction of which both numerator and denominator do not vanish, for the required value of x; this fraction is the value required.*

We have already shown that we must arrive at such a fraction.

5. Required the value of $\frac{(a-x)^4}{(3a-3x)^3}$ when $x = a$.

$$\text{Ans. } \frac{p'''}{q'''} = \frac{-24(a-x)}{-162} = \frac{0}{162} = 0.$$

6. Required the value of $\frac{3x^2-3}{4x^3-12x+8}$ when $x = 1$.

Ans. ∞, by 1st dif. coefs.

7. Required the value of $\frac{x^3-a^3}{x^2-a^2}$ when $x = a$.

Ans. $3a$, by 2d dif. coefs.

8. Required the value of $\frac{x+x^2-2}{(1-x)^3}$ when $x=1$.

Ans. $-\frac{1}{3}$ by 2d dif. coefs.

9. Required the value of $\frac{x^3-3x+2}{x^4-6x^2+8x-3}$ when $x=1$.

Ans. ∞ by 2d dif. coefs.

173. Inasmuch as we have deduced the process for finding the value of $\frac{Fx}{fx}$ when it becomes $\frac{0}{0}$, we have, in effect, found the process for finding the value of $\frac{Fx}{1} \times \frac{1}{fx}$ (of which expression the first factor is certainly 0, and the second is ∞), that is, for finding the value of a product of two functions as factors, one of which becomes 0 and the other ∞ when the variable takes a particular value. We have only to take the first as a numerator and the reciprocal of the second as denominator and we have the vanishing fraction just investigated.

10. Required the value of $(x^n - 1) \times \frac{1}{x-1}$ when $x=1$.

Ans. n.

174. The foregoing demonstration of the process for finding the value of a vanishing fraction embraces the principle of finding the value of a fraction which becomes $\frac{\infty}{\infty}$ under the same condition; for any fraction is the same in value as the reciprocal of its denominator taken for numerator, and the reciprocal of its numerator taken for the denominator. Thus, if

$$\frac{Fx}{fx} = \frac{\infty}{\infty},$$

then

$$\frac{Fx}{fx} = \frac{\frac{Fx}{Fx\,fx}}{\frac{fx}{Fx\,fx}} = \frac{\frac{1}{fx}}{\frac{1}{Fx}} = \frac{0}{0}.$$

We remark that evidently the reciprocal of an infinite quantity is zero, i. e., $\frac{1}{\infty} = 0$.

11. Required the value of $\frac{50}{a - x} \div \frac{20}{a^2 - x^2}$ when $x = a$.

Ans. $5\,a$.

175. Lastly, the demonstration also embraces the case of determining the value of the difference of two functions, each of which, for a particular value of the variable, becomes infinite; for, in subtraction, any remainder is equal the fraction of which the numerator is the excess of the reciprocal of the subtrahend above the reciprocal of the minuend, and the denominator is the reciprocal of the product of minuend and subtrahend. Thus, Fx being ∞, and fx being ∞,

$$Fx - fx = \frac{Fx - fx}{1} = \frac{\frac{Fx - fx}{Fx \times fx}}{\frac{1}{Fx \times fx}} = \frac{\frac{1}{fx} - \frac{1}{Fx}}{\frac{1}{Fx \times fx}} = \frac{0}{0}.$$

11. Required the value of $\frac{1}{x^3 - a^3} - \frac{b}{x - a}$ when $x = a$.

Ans. ∞.

176. Whenever an infinite quantity is generated by the denominator of a fraction becoming 0, since $-0 = +0$, it is evident that such infinite quantity has the ambiguous sign $\pm$, and becomes $\pm \infty$.

177. The same characteristics of different values belong to infinite quantities that belong to finite quantities and to zero, dependent upon their mode of generation, or of relation to each other by factors, by radical expressions, or otherwise. It is not considered that a finite quantity is any addition to an infinite one, or diminution from one, and such finite quantity may be expunged.

178. An infinite quantity is an impossible quantity; hence all conclusions predicated on the possibility of an infinite quantity must fail; with the exception, however, that certain conclusions are practicable with reference to the finite terms of a series. But general developments fail for certain values of a quantity when any term of a series becomes infinite for such values.

SECTION XXII.

EXCEPTIONAL PRINCIPLE RELATING TO TAYLOR'S THEOREM.

179. The general development of every function of a variable according to the ascending entire and positive powers of its increment, for *general values* of the variable, is possible by Taylor's Theorem. But this development does not hold whenever, for a *particular value* of the variable, any of the coefficients of Taylor's series become infinite. Thus, the general development of $F'x = \sqrt{(x - a)}$, when x is replaced by $x + h$, is, by this theorem,

$$F'(x + h - a) =$$

$$(x + a)^{\frac{1}{2}} + \frac{1}{2}(x - a)^{-\frac{1}{2}} h - \frac{1}{8}(x - a)^{-\frac{3}{2}} h^2 + \text{etc.},$$

where, in case $x = a$, all the differential coefficients become infinite. And it will be observed that when any dif. coef. becomes infinite, all succeeding ones do also.

It is not held that the development fails when these coefficients become imaginary if the variable takes a particular value; because the function of $x + h$ would itself become imaginary at the same value, and it is proper that one imaginary quantity should be equated with another.

Since such coefficients as become infinite for the proposed value of the variable, become so on the principle of the denominator of a fraction vanishing, the sign of such infinite coefficient becomes always $\pm$, or is ambiguous.

Whatever uses, therefore, we may on general principles wish to make of Taylor's Theorem, become exceptional on the condition alluded to. Thus, in regard to maxima and minima, suppose it were

1. Required to find the maxima or minima values of the function $y = b + (x - a)^{\frac{4}{3}}$.

$$\therefore \frac{dy}{dx} = \frac{4}{3}(x-a)^{\frac{1}{3}},$$

$$\frac{d^2 y}{dx^2} = \frac{4}{9}(x-a)^{-\frac{2}{3}}.$$

The equation $\frac{dy}{dx} = 0$ gives $x = a$; and if we proceed to determine whether we have a maximum or minimum for $x = a$, we find that for this value the second dif. coef. is infinite, which is but another name for an impossible quantity. But as we have never any thing to do with the *greatness* of a dif. coef., when we examine it with the purpose we now have in view, but have to do with its *sign* only, this second dif. coef. must, from its mode of deriving its infinite value, have the ambiguous sign $\pm$; therefore the function in question must, at $x = a$, be inferred to have both a maximum and a minimum, which nevertheless is still a possibility for some functions, but, with reference to the one in question, may be found by algebraic or arithmetical tests not to be true, but that there is only a minimum. For if we test the value of y immediately before $x = a$, as when $x = a - h$, and immediately after $x = a$, as when $x = a + h$, that is, by substituting $a \pm h$ for x in the function, we shall have

$$F(a \pm h) = b + h^{\frac{4}{3}}.$$

Hence b is increased for either sign of h; consequently $x = a$ renders the function a minimum when having the value b.

It will be observed that since the *odd* root of a negative quantity is possible, such root being negative, and the *even* power of all quantities is positive, the fourth power of the third root is positive.

180. To obtain the true development of a function for that one, or those, particular values of the variable which cause Taylor's Theorem to fail, the usual course is to recur to the ordinary process of common algebra, after having substituted $a + h$ for x in $F\,x$.

2. Required the development of $F\,x = 2\,a\,x - x^2 + a\sqrt{x^2 - a^2}$ for the condition when x becomes $a + h$; and to arrange the terms according to the increasing exponents of h.

Substituting $a + h$ for x we have

$$F\,(a + h) = a^2 - h^2 + a\,h^{\frac{1}{2}}\,(2\,a + h)^{\frac{1}{2}},$$

developing the binomial $(2\,a + h)^{\frac{1}{2}}$ by the Binomial Theorem, and multiplying its terms by $a\,h^{\frac{1}{2}}$, we have

$$F\,(a + h) = a^2 + a\,(2\,a)^{\frac{1}{2}}\,h^{\frac{1}{2}} + \frac{a\,h^{\frac{3}{2}}}{2\,(2\,a)^{\frac{1}{2}}} - h^2 - \frac{a\,h^{\frac{5}{2}}}{8\,(2\,a)^{\frac{3}{2}}} + \text{etc.}$$

The algebraic process in question is any that will reduce complex terms to simple ones, in which h shall appear as a factor with any whole or fractional index, unless, perhaps, it may as above be eliminated from any term or terms. It will be observed that these various coefficients of h are not *differential* coefficients.

3. Required to determine whether y in the following $F\,(x, y) = 0$ has a maximum or minimum, viz.:

$$(y-b)^3=(x-a)^2.$$

$$\therefore\ \frac{dy}{dx}=\frac{2}{3}\cdot\frac{x-a}{(y-b)^2}=\frac{2}{3}\,\frac{1}{(x-a)^{\frac{1}{3}}}.$$

In this instance we might, on inspecting its first form, incautiously infer that $\frac{dy}{dx}$ becomes 0 when $x=a$, if we regard only the numerator (Art. **112**). But we are obliged to inquire whether we have not here a vanishing fraction, the denominator becoming 0 when $x=a$, which we should find to be true, and that its value is infinite when $x=a$; which we see at once on inspecting the second form. And we find that x is infinite, by the rule of vanishing fractions, when the first dif. coef. $=0$, and therefore (Art. **108**) we should have no maximum or minimum. If, nevertheless, we consider the infinite value of the first dif. coef., we find it occurs when $x=a$, and at this value Taylor's Theorem fails. Yet if we test this value, $x=a$, in the function, or rather the values of x, within h of a, we shall find a minimum. Substituting $a\pm h$ for x we have for y

$$y=b+h^{\frac{2}{3}},$$

and b is increased for either sign of h, since all possible values of y are b and something additional to b; hence a minimum. Hence an important principle supplementary to our Section on Maxima and Minima, which is:

181. Before we can conclude in any case that the values of x deduced from the condition $\frac{dy}{dx}=0$ comprise among them all those that can render a function a maximum or minimum, we must examine those values of x arising from the condition $\frac{dy}{dx}=\infty$. And as this is a case where the development by Taylor's Theorem fails, we must make this examination by the algebraic method of substituting each of these values, as affected by the addition and subtraction of h, for x, in the proposed function, and observing which of the results agree with the conditions of maxima and minima, as by definition.

182. The method pointed out in a previous section for determining the value of a vanishing fraction now requires the mention, that in case a numerator or denominator fails to be developable by Taylor's Theorem, we must adopt the algebraic method of development, for either or both which so fail. We should then arrange the terms as numerator and denominator according to the increasing exponents of h; then divide each term in either by h at the lowest power of either; then test what the fraction becomes for $h = 0$. Whatever value it has is the value desired.

4. Required the value of $\frac{(x^2 - a^2)^{\frac{3}{2}}}{(x - a)^{\frac{3}{2}}}$ when $x = a$.

$$\frac{F(a+h)}{f(a+h)} = \frac{(2ah + h^2)^{\frac{3}{2}}}{h^{\frac{3}{2}}} = \frac{(2ah)^{\frac{3}{2}} + \frac{3}{2}(2ah)^{\frac{1}{2}} h^4 +, \text{ etc.}}{h^{\frac{3}{2}}} =$$

$$\frac{(2a)^{\frac{3}{2}}}{1} = (2a)^{\frac{3}{2}}, \quad \text{Ans.}$$

5. Required $\frac{dy}{dx}$ in $y = x + (x - a)^2 \sqrt{x}$ for $x = a$.

Ans. 1.

6. Required $\frac{d^2y}{dx^2}$ in $y = x + (x - a)^2 \sqrt{x}$ for $x = a$.

Ans. $\pm 2\sqrt{a}$.

7. Required $\frac{dy}{dx}$ and $\frac{d^2y}{dx^2}$ in $(y - x)^2 = (x - a)^4 x$ for $x = a$.

Ans. $\frac{dy}{dx} = 1$, $\frac{d^2y}{dx^2} = \pm 2\sqrt{a}$.

It is worthy of notice whether the last implicit function is not virtually the preceding explicit one when we regard y.

8. Required $\frac{dy}{dx}$ in $y^3 = (x - a)^3 (x - b)$ for $x = a$.

Ans. $(a - b)^{\frac{1}{3}}$.

We here terminate the portion of our treatise relating strictly to Algebraic Functions.

SECTION XXIII.

NATURE OF LOGARITHMS AND EXPONENTIAL QUANTITIES.

183. Thus far we have made no mention of any functions in connection with which the variable occurs as an exponent, whether of a power, as in a^x, or of a root, as in $b^{\frac{1}{x}}$, or of a power and a root which is the characteristic of a fraction in general as exponent. The reason of this has been a regard to a distinctive division of subjects. The discussion of the properties and differentiation of strictly algebraic functions being completed, we shall be brought to consider, in the succeeding section, functions of a new order. But we are not to forget that the subjects strictly relate to numerical analysis, and are directly consecutive with our preceding inquiries. It is, however, called *transcendental* analysis, as indicative of being of a grade above what is commonly called algebraic.

Since many treatises of algebra, otherwise quite complete, do not contain an account of the theory and uses of logarithms, and it would be unfortunate for us to go forward without such preparation, we will devote this section to the Nature of Logarithms and Exponential Quantities.

184. A logarithm is such exponent as, applied to any number a greater than 1, shall cause a number to be denoted equal to any positive numerical quantity b whatever, greater or less than 1. The logarithm in question is the logarithm of b, the latter number.

185. A *system* of logarithms is a collection of exponents such as offer by selection, one which, when applied to some constant number (originally arbitrary), called the *base*, will render this base the equal and the representative of any number whatsoever; and hence every number whatsoever. This exponent is the logarithm of the latter number.

186. In accommodation to the decimal system of numbers, the number 10 has been selected as the base of the *common system.* Accordingly, in this system 1 is the logarithm of 10, because $10^1 = 10$; 2 is the logarithm of 100, because $10^2 = 100$; 3 of 1000, because $10^3 = 1000$, etc. Zero or 0 is the logarithm of 1, because $10^0 = 1$. As it is seemingly arbitrary to call 0 the common logarithm of 1, we remark that it is strictly inferred from the ratio by which logarithms diminish with entire units. We have for logarithms, by continuity, the following, placed in connection with the numbers of which they are the logarithms:

3	2	1	0	-1	-2	-3,	etc.
1000	100	10	1	$\frac{1}{10}$	$\frac{1}{100}$	$\frac{1}{1000}$,	etc.
or,				,1	,01	,001 ,	etc.

whence it appears that while all the *natural numbers*,

when selected for their logarithms in entire numbers, vary successively by the ratio 10 or $\frac{1}{10}$, their logarithms vary by a uniform difference of 1. And it appears that we pass zero with the preservation of this law.

It already appears that the logarithm of a fraction, by which we mean a numerical quantity less than 1, is negative. If we represent the logarithm of 0 by $-n$, we have

$$10^{-n} = 0 = \frac{1}{10^n},$$

which requires that the denominator 10^n be infinite, or what is the same, n to be infinite.

It further appears that the common logarithm of any number greater than 1 and less than 10 must be between 0 and 1, i. e., be a fraction, or rather we should find it not to be expressible exactly even as a fraction. The same remark applies to any logarithm which is not a whole number, positive or negative; that is, the logarithm of any number intervening between 1000 and 100; 100 and 10; 10 and 1; 1 and ,1; ,1 and ,01; ,01 and ,001; etc., is expressed by a whole number, or 0, and a fraction, most conveniently a decimal fraction. Such decimal fraction not absolutely expressing the value intended, is what is known as an irrational quantity. Thus, the common logarithm of 2 is 0,3010299, that of 543 is 2,7352793, etc. In view of this incommensurableness of most numbers and their respective logarithms, only an approximate definition can be given of a logarithm in general. The definition should embrace a reference to a certain power of a certain root of the understood base, by successive degrees of proposed approximation, each replacing the preceding. Thus the common logarithm of 2 is, by the degree of nearness desired,

$$\frac{3}{10}, \quad \text{or} \quad \frac{301}{1000}, \quad \text{or} \quad \frac{30102}{100000}, \quad \text{or} \quad \frac{301029}{1000000}, \quad \text{etc.};$$

which are to be read, 10th root of the 3d power of the base, or the 1000th root of the 301st power, etc. In this way must we enunciate the significance of a decimal fraction as an exponent. In proper mathematical expression, as we shall see, logarithms, when not entire numbers, are discoverable as existing in an infinite series, which indeed a decimal fraction not terminating, itself is.

The method of using logarithms for the purpose of facilitating operations with numbers is quite evident after the study of algebraic exponential quantities.

187. In order to multiply quantities, we add their logarithms; the sum of their logarithms is the logarithm of the product, or continued product, of two or more quantities.

Hence any number is multiplied by 10 by adding 1, the common logarithm of 10, to that of the number; this logarithm, thus increased, becomes the logarithm of the product. It is multiplied by 100 by adding 2, by 1000 by adding 3 to its logarithm. Advantage is taken of this property in the preparation of tables to insert only the fractional part of a logarithm, leaving the integral part, or *characteristic*, to be extemporized according to (being one less than) the number of places of the *integral part* of the number of which the logarithm is desired. Thus, the logarithm of

54360	is	4,7352794
5436	is	3,7352794
543,6	is	2,7352794
54,36	is	1,7352794
5,436	is	0,7352794
,5436	is	$\bar{1}$,7352794

The logarithm of ,5436 is expressed here with its decimal part positive, while its integral part is negative. The practice is in common use of adding 10 to the characteristic in such cases, since the inconvenience of the negative characteristic is thus avoided, and no error would be likely to arise in common uses which would not be strikingly obvious, and easily corrected by subtracting the 10. Thus the logarithm of ,5436 is expressed as 9,7352794.

188. Division of numbers, being the converse of multiplication, is effected by the subtraction of the logarithm of one number from that of the other; this difference, when positive, is the logarithm of the number of times the less is contained in a greater number; when negative, it is the logarithm of the fractional time the greater is contained in the less. We have already found that the logarithm of a fraction is negative.

189. In order to raise a numerical quantity to any power, we multiply the logarithm of that quantity by the number denoting the power required; the product is the logarithm of the power required.

190. In order to extract any root of a numerical quantity, we divide its logarithm by the cardinal number expressing the root required in the ordinal form of expression, as 2 for second, etc.

The method of finding, by the use of logarithms, the fourth term of a set of common direct proportionals, is therefore extremely obvious; we add the logarithms of the first and second terms; from the sum subtract that of the third; the remainder is the logarithm of the fourth, or term required.

A small volume is to be obtained containing a table of common logarithms, for all numbers from 0 to 10000, to six places of decimals, sometimes to seven places. The method of taking out the logarithm for any number within these limits, and of extending the use of the table to much greater numbers, as well as of finding the natural number corresponding to any possessed logarithm is usually printed with the table. The table is also to be found in treatises of navigation and surveying.

The logarithm of a negative quantity does not belong to the same system with those of positive quantities. When, however, certain numerical operations with negative quantities are to be done, we may eliminate the condition of their negativeness until the result is reached, when the appropriate sign may be prefixed to it, as algebraically determined.

SECTION XXIV.

DIFFERENTIATION AND DEVELOPMENT OF LOGARITHMIC AND EXPONENTIAL FUNCTIONS.

191. A logarithmic function is the logarithm of a variable quantity; as, log. x, or log. $(b + x^n)$, which do not denote the logarithm which is x, etc., but the logarithm of the number which is x, etc.

192. An exponential function is one in which the variable, or some function of it, holds the position of index or exponent; as, a^x, or b^{nx}, the root being a constant.

It is important to observe that such index, when considered as some logarithm, is not the logarithm of the *same quantity* to which it is attached as index, but of the entire power which itself is employed in expressing.

A number or numerical quantity, and its logarithm, are distinguished as *natural number*, or quantity, and its logarithm.

193. We now proceed to differentiate

$$y = \log. x,$$

for any system of logarithms having for its base a, which convenience will require to be considered greater than 1; for we immediately derive, by the converse of the definition of a logarithm,

$$x = a^y, \qquad (1.)$$

and by no variation of y while positive can a^y represent all numerical quantities, unless a be greater than 1.

Letting y take the increment h as the independent variable, and x' denote the corresponding value of x, we have

$$x' = a^{y+h} = a^y \times a^h.$$

Let us now substitute $1+b$ for a, and develop $(1+b)^h$ by the binominal theorem, and we have

$$a^h = (1+b)^h = 1 + hb + \frac{h}{1}\cdot\frac{h-1}{} b^2 +$$

$$\frac{h}{1}\cdot\frac{h-1}{2}\cdot\frac{h-2}{3} b^3 +, \text{etc.}$$

The multiplication of $h\,(h-1)\,(h-2)$, etc., being done, and all the quantities selected from the successive terms of the continued series, which are factor to h, and placed or indicated within the following parenthesis, and $s\,h^2$ being used for all succeeding terms, in which s alone includes, in some sense, h, or a series involving h, we have

$$a^h = 1 + h\,(b - \tfrac{1}{2}b^2 + \tfrac{1}{3}b^3 - \tfrac{1}{4}b^4 +, \text{etc.}) + s\,h^2;$$

multiplying both members by a^y, and calling the terms within the parenthesis c, we have

$$a^y \times a^h = x' = (1 + ch + sh^2)\, a^y;$$

Subtracting the equals $x = a^y$, from the above equals, we have

$$x' - x = c\,a^y h + s\,a^y h^2$$

$$\therefore \frac{x' - x}{h} = c\,a^y + s\,a^y h,$$

which becomes, when $h = 0$,

$$\frac{dx}{dy} = c\,a^y. \qquad (2.)$$

Now, a^y being $= x$ by (1), we have

$$\frac{dx}{dy} = cx;$$

$$\text{i. e.} \quad \frac{dy}{dx} = \frac{1}{c} \times \frac{1}{x},$$

$$\text{and} \quad dy = \frac{1}{c} \times \frac{dx}{x}. \qquad (3.)$$

where b being $= a - 1$, we have

$$\frac{1}{c} = \frac{1}{a - 1 - \frac{1}{2}(a-1)^2 + \frac{1}{3}(a-1)^3 - \frac{1}{4}(a-1)^4 + \text{etc.}}$$

Defining now $\frac{1}{c}$, or the reciprocal of c, as the *modulus* of the system of logarithms which has the number a for its base, we have the general rule:

194. *To differentiate a logarithmic function, or the logarithm of a variable quantity, we must multiply the modulus of the system by the differential of the natural quantity, and divide the product by the natural quantity itself.*

195. Since a logarithmic function may be of a more complex form than simply log. x, some function of x being in the place of x, we must evidently have in the result, in the place of $d\,x$, the whole differential of such function, in which, indeed, $d\,x$ will be found as a factor.

196. If the variable quantity of which the logarithm is intended in the constitution of a logarithmic function, *contain a constant factor*, since such factor will be found, in pursuance of the rule for differentiation, as still a factor of the numerator and of the denominator, it becomes evident that such factor contributes nothing affecting the differential. This is consistent with a previous change of the function into the sum of the logarithms of the factors; thus:

$$d\,[\log.\,b\,(a-x)\,] = d\,[\log.\,b + \log.\,(a-x)\,] =$$

$$d\,[\log.\,(a-x)\,] = -\frac{d\,x}{a-x}\cdot\frac{1}{c}.$$

197. But if the logarithmic function be associated with a factor in a manner by which the logarithm of such product is not intended, the constant factor will be found affecting the differential as factor; thus,

$$d\,[m\log.\,x] = m\,d\log.\,x = \frac{m\,d\,x}{x}\cdot\frac{1}{c}$$

198. Selecting from the demonstration of the differentiation of logarithmic functions, equation (1), viz.,

$$x = a^{y}$$

we observe a^y to be an exponential function, and that its differential coefficient is expressed in equation (2), y being the independent variable, and x being dependent. It is

$$\frac{d\,x}{d\,y} = c\,a^{y}$$

$$\therefore d\,x = c\,a^{y}\,d\,y.$$

199. *Therefore, to differentiate an exponential function, we must multiply together the reciprocal of the modulus of the system of logarithms, determined by the base of the exponential, the exponential itself, and the differential of the variable exponent.*

200. An exponential function is not considered to be restricted to the very simple form of a^y, or indeed to be restricted at all, for b^{ay}, $a^y + b^{na}$, $\frac{1}{(a^y c)^y} + b^y$, etc., are held to be examples; the *base* must be, however, the root of the power indicated, and in $b\, a^y$, b is no part of the base, nor even a factor of it, but is a factor of the power only.

If in the preceding demonstration we had at the outset,

$$x = b\, a^y,$$

we should find the factor, b, passing through the demonstration, and appearing in the result,

$$dx = c\, b\, a^y\, dy;$$

in the sequel we shall have a practical use for this observation.

201. For the hyperbolic or Napierian system of logarithms, the modulus has been assumed $= 1$, which value renders its reciprocal $= 1$. As to this system, therefore, the mention of the modulus may be eliminated from the two preceding italicized rules; if in the succeeding context all mention of a modulus is omitted in any operation, the hyperbolic system will be understood to be intended.

202. For $c = 1$, the value of a, as the base of the hyperbolic system, must be deduced. We will therefore develop, by Maclaurin's Theorem, the exponential function a^y according to the powers of y.

$$\text{Let } x = a^y \therefore \text{ when } y = 0, x = 1;$$

$$\frac{dx}{dy} = c\,a^y \therefore \text{ when } y = 0, x = c;$$

$$\frac{d^2x}{dy^2} = c^2 a^y \therefore \text{ " \quad " \quad " } x = c^2;$$

$$\frac{d^3x}{dy^3} = c^3 a^y \therefore \text{ " \quad " \quad " } x = c^3;$$

etc. etc.

$$\because a^y = 1 + cy + \frac{c^2y^2}{1.2} + \frac{c^3y^3}{2.3} +, \text{ etc.}$$

When, therefore, $c = 1$ and $y = 1$, we have

$$a - 1 + 1 + \frac{1}{2} + \frac{1}{2.3} + \frac{1}{2.3.4} +, \text{ etc.}$$

$$= 2, 71828,$$

which is the base of the hyperbolic system.

But if we wish to assume $a = 10$, which is very desirable for, and is the base of the common system, since $a - 1 = 9$, we have for c

$$c = 9 - \tfrac{1}{2}(9)^2 + \tfrac{1}{3}(9)^3 - \tfrac{1}{4}(9)^4 +, \text{ etc.}$$

$$= 2, 30258509;$$

and $\frac{1}{c} =$,43429448, which is the modulus of the common system.

203. For the common system of logarithms, therefore, the fraction ,43429448, must be read as modulus in the foregoing rules for the differentiation of logarithmic functions, (Art. **194**); and the number, 2,30258509 as the reciprocal of the modulus in the rule for the differentiation of exponential functions (Art. **199.**).

To recapitulate: we have for

Modulus of hyp. system, assumed, . . .	1,
Base of the hyp. system deduced, . . .	2,71828
Reciprocal of this modulus, ...	1,

Base of the com. system assumed, . .	10,
Modulus of com. system deduced, . . .	,43429448
Reciprocal of this modulus, ...	2,30258509

Since from equation (3) foregoing, which is

$$dy = \frac{1}{c} \times \frac{dx}{x},$$

we derive

$$\frac{dy}{dx} = \frac{1}{c} \times \frac{1}{x},$$

which becomes, when $x = 1$,

$$\frac{dy}{dx} = \frac{1}{c},$$

we have for the dif. coef. of the logarithm of **1** in every system, the modulus of such system. The modulus of a system is therefore the ratio, or rate, at which positive logarithms come into being. And this ratio is constant for whatever number x may be. Accordingly, whatever the number may be, the ratio of its logarithms, by different systems, is always constant. Hence, we may find the hyperbolic logarithm of any number from its common logarithm, by multiplying the latter by 2,30258509. And, conversely, a hyperbolic logarithm may be converted into a common logarithm by dividing it by 2,30258509.

In regard to negative logarithms, or those of fractions, or any numerical quantities less than 1, the pursuance of this multiplication renders the hyperbolic the *less* logarithm than the common, the greater negative being, of course, the less quantity.

For the most concise method of indicating whether the hyperbolic or common logarithm is intended, authors agree to cite the hyperbolic, by the small Roman letter l., or log., the common, by the Roman capital L., or Log. *We shall adopt this distinction* hereafter, when distinction is necessary.

Mr. J. R. Young suggests that $(\log.)^2 x$ shall be taken to signify log. log. x, or logarithm of the logarithm of x, but $\log.^2 x$, having no parenthesis, to signify the second power of log. x. Sufficiently explicit is log. x^2 for the logarithm of x^2. We will adopt this use.

When $\frac{a}{b}$ is a quantity less than one, log. $\frac{a}{b}$ is negative without the expression by a negative sign; hence $\left(-\log. \frac{a}{b}\right)$ becomes a positive quantity; hence log. $\left(-\log. \frac{a}{b}\right)$ is the logarithm of a positive quantity, and becomes entitled to the abridgment, $-(\log.)^2 \frac{a}{b}$. The succeeding context presents a case of this use.

SECTION XXV.

EXAMPLES OF DIFFERENTIATION AND DEVELOPMENT OF LOGARITHMIC FUNCTIONS.

1. Required to develop log. $(a+x)$ by Maclaurin's Theorem, according to the ascending powers of x.

Let $y = \log.(a+x) \therefore \quad (y) = \log. a,$

$$\frac{dy}{dx} = \frac{1}{a+x} \qquad \therefore \left(\frac{dy}{dx}\right) = \frac{1}{a},$$

$$\frac{d^2y}{dx^2} = -\frac{1}{(a+x)^2} \qquad \therefore \left(\frac{d^2y}{dx^2}\right) = -\frac{1}{a^2},$$

$$\frac{d^3y}{dx^3} = \frac{2}{(a+x)^3} \qquad \therefore \left(\frac{d^3y}{dx^3}\right) = \frac{2}{a^3},$$

$$\frac{d^4y}{dx^4} = -\frac{2.3}{(a+x)^4} \qquad \therefore \left(\frac{d^4y}{dx^4}\right) = -\frac{2.3}{a^4},$$

$$\therefore \log.(a+x) = \log. a + \frac{x}{a} - \frac{x^2}{2a^2} + \frac{x^3}{3a^3} - \frac{x^4}{4a^4} +, \text{ etc.}$$

2. From the above development, required to determine the Log. of the number 11.

$$\log.(10+1) = \log. 10 + \frac{1}{10} - \frac{1}{2.10^2} + \frac{1}{3.10^3} - \frac{1}{4.10^4} +, \text{ etc.}$$

$$\log. 11 = 2{,}30258 + \frac{1}{10} - \frac{1}{200} + \frac{1}{3000} - \frac{1}{40000} +, \text{ etc.}$$

$$= 2{,}39788$$

$$\therefore 2{,}39788 \div 2{,}30258 \doteq 1{,}04139. \quad \text{Ans.}$$

The algebraic addition of six terms only of the above series for log. 11, is sufficient for determining the hyp. log.

correctly to five places of decimals. This result, divided by the reciprocal of the modulus of the common system, gives the common logarithm of 11.

If a be quite large, as 4000, or 5000, the addition of only two terms, viz., log. 5000 and $\frac{1}{5000}$ the dif. coef. of log. $(5000+1)$, gives log. 5001, accurately to five places of decimals.

We can never call the "differences" between the logarithm of one number, and that of a number greater by 1, the differential of the logarithm of that number; the differential of a logarithmic function is 0, and not 1.

3. Required the development of $y = \log. (1+x)$ for $x=1$.

$$\text{Ans.} \quad y = \log. 2 = 1 - \frac{1}{2} + \frac{1}{3} - \frac{1}{4} + \frac{1}{5} - \frac{1}{6} +, \text{ etc.}$$

The summation of this series would serve to give the hyp. log. of 2, were it not of such slow convergency as to render a correct sum for six places exceedingly laborious.

4. Required Log. 2 from the development of $y = \log. \frac{1+x}{1-x}$, which, when $x = \frac{1}{3}$, becomes evidently log. 2.

$$\therefore (y) = \log. 1 = 0;$$

$$\frac{dy}{dx} = \frac{2}{(1-x^2)^2} \div \frac{1+x}{1-x} \therefore \left(\frac{dy}{dx}\right) = 2 = \frac{2-2x}{1-x^2};$$

$$\frac{d^2y}{dx^2} = \frac{4x}{(1-x^2)^2} \therefore \left(\frac{d^2y}{dx^2}\right) = 0;$$

$$\frac{d^3y}{dx^3} = \frac{4(1-x^2)^2 - 4x(-4x+4x^3)}{(1-x^4)^4} \therefore \left(\frac{d^3y}{dx^3}\right) = 4, \text{ etc.};$$

$$\therefore \log. \frac{1+x}{1-x} = 2x + \frac{2}{3}x^3 + \frac{2}{5}x^5 + \frac{2}{7}x^7 +, \text{ etc.}$$

We have, then, if $x = \frac{1}{3}$,

$$\begin{aligned}
2x &= 0{,}66666666 \\
\frac{2}{3}x^3 &= \phantom{0{,}6}2469134 \\
\frac{2}{5}x^5 &= \phantom{0{,}66}164614 \\
\frac{2}{7}x^7 &= \phantom{0{,}666}13064 \\
\frac{2}{9}x^9 &= \phantom{0{,}6666}1128 \\
\frac{2}{11}x^{11} &= \phantom{0{,}66666}102
\end{aligned}$$

$\therefore$ log. 2 = ,69314708

Now ,69314708 ÷ 2,30258 = ,30103 $\therefore$,30103 is Log. 2.

Having obtained Log. 11 = 1,04139, and Log. 2 = ,30103, we find

$$\begin{aligned}
&1{,}04139 + {,}30103 && = \text{L. } 22 && = 1{,}34242 \\
&1{,}34242 + 1 && = \text{L. } 220 && = 2{,}34242 \\
&1{,}04139 - {,}30103 && = \text{L. } 5\tfrac{1}{2} && = 0{,}74036 \\
&1{,}04139 \times 2 = \text{L. } 11^2 && = \text{L. } 121 && = 2{,}08278 \\
&0{,}74036 + 1 && = \text{L. } 55 && = 1{,}74036 \\
&1{,}74036 \times 2 = \text{L. } 55^2 && = \text{L. } 2925 && = 3{,}48072, \text{ etc.}
\end{aligned}$$

When we have obtained the logarithms of the prime numbers, we easily, as above, obtain the logarithms of all other numbers.

5. Required to differentiate $y = x \log. x, = \log. x^x$.

Ans. $dy = dx \log. x + x \times \frac{dx}{x} = dx \log. x + dx.$

6. Required to differentiate $y = \log. x^2 = 2 \log. x$.

Ans. $dy = \frac{2\,dx}{x}.$

7. Required to differentiate $y = \log. x + \log. (a + x) = \log. [x \times (a + x)] = \log. (a x + x^2)$.

Ans. $d y = \frac{(a + 2x) d x}{a x + x^2}$.

8. Required to differentiate $y = \log.^2 x$.

Making $\log. x = z, d y = 2 z d z = 2 \log. x d z = \frac{2 \log. x \times d x}{x}$.

9. From $\log. y = x$, to find $d y$.

Ans. $\frac{d y}{y} = d x \therefore d y = y d x$.

10. From $a = \log. x^y$, to find $d y$.

$$\therefore a = y \log. x, \therefore y = \frac{a}{\log. x} \therefore d y = -\frac{a d \log. x}{\log.^2 x}, = -\frac{a d x}{x \log.^2 x}.$$

11. Required to differentiate $y = (\log.)^2 x$, by which is intended $\log. \log. x$.

Putting z for $\log. x$, $d y = \frac{d z}{z}$, but $d z = d \log. x$

$$\therefore dy = \frac{d x}{x \log. x}.$$

12. Required the value of the following vanishing fraction, when $x = 1$, or, which is the same, the value of the difference of the two functions of x, each of which becomes ∞ when $x = 1$, viz.:

$$\frac{x \log. x - (x - 1)}{(x - 1) \log. x} = \frac{x}{x - 1} - \frac{1}{\log. x}.$$

Art. **171.** $\frac{p'}{q'} = \frac{\log. x + 1 - 1}{\log. x + \frac{x - 1}{x}}, \frac{p''}{q''} = \frac{\frac{1}{x}}{\frac{1}{x} + \frac{x - x + 1}{x^2}} = \frac{1}{2}.$

13. Required the value of $\frac{\log. x - x \log. x}{(\log. x)^2}$ when $x = 1$.

Ans. -1.

204. The differentiation of an algebraic function, which is resolvable into factors, may be much facilitated by first taking its logarithm, and then differentiating, it being observed that the differential of a logarithmic function may not contain a logarithm, but be purely algebraic. If the function is not algebraic, and is not resolvable into factors, this method may be used, but without advantage.

14. Required $\frac{dy}{dx}$ of $y = (a + x^2)^3 \sqrt{x}$.

We have $\log. y = 3 \log. (a + x^2) + \frac{1}{2} \log. x$

$$\therefore d \log. y = \frac{dy}{y} = \frac{6 x dx}{a + x^2} + \frac{dx}{2x}. \quad (1.)$$

$$\therefore \frac{dy}{dx} = 6 (a + x^2)^{\frac{5}{2}} + \frac{1}{2} x^{-\frac{1}{2}} (a + x)^3,$$

after substituting the value of y for y in equation (1.)

15. Required $\frac{dy}{dx}$ of $y = x (a^2 + x^2) \sqrt{a^2 - x^2}$,

$$\therefore \log. y = \log. x + \log. (a^2 + x^2) + \frac{1}{2} \log. (a^2 - x^2)$$

$$\therefore d \log. y = \frac{dy}{y} = \frac{dx}{x} + \frac{2x dx}{a^2 + x^2} - \frac{x dx}{a^2 - x^2},$$

$$= \frac{a^4 + a^2 x^2 - 4 x^4}{x (a^2 + x^2)(a^2 - x^2)} dx$$

$$\therefore \frac{dy}{dx} = \frac{a^4 + a^2 x^2 - 4 x^4}{(a^2 - x^2)^{\frac{1}{2}}}.$$

205. Since the logarithm of a quantity or function becomes greater as the quantity becomes greater, and less as the quantity becomes less, the maximum of the quantity occurs at the same value of the variable, as the maximum of its logarithm occurs. We may avail ourselves of this principle.

Problem 33, on page 88, rendered in more general terms, becomes:

16. Required to divide the number a into two such parts that the mth power of one part multiplied by the nth power of the other, shall be a maximum.

Let $x =$ one of the parts, and $y =$ the product in question; then,

$$y = x^m (a - x)^n,$$

let $$u = \log. y = m \log. x + n \log. (a - x),$$

then $$d u = d \log. y = \frac{m\,d x}{x} - \frac{n\,d x}{a - x} = \frac{d y}{y}$$

$$\therefore \frac{d u}{d x} = \frac{d \log. y}{d x} = \frac{m y}{x} - \frac{n y}{a - x} = \frac{d y}{y\,d x}.$$

Whence, if $\frac{d u}{d x} = 0$, or $\frac{d y}{y\,d x} = 0$, we have $x = \frac{m a}{m + n}$.

We find that we thus eliminate all necessity of substituting the value of y in $\frac{d u}{d x}$, or $\frac{d y}{y\,d x}$.

We will now prove that we need not substitute the value of y in the second or any succeeding differentiation of $\frac{d u}{d x}$, so far as determining maxima is concerned.

We have $$\frac{y\,d\,u}{d\,x}=\frac{d\,y}{d\,x},$$

$$\therefore\ \frac{d\,x\,(d\,y\,d\,u+y\,d^2\,u)}{d\,x^2}=\frac{d^2\,y}{d\,x^2};$$

therefore, when $$\frac{d\,u}{d\,x}=0,$$

$$\frac{y\,d^2\,u}{d\,x^2}=\frac{d^2\,y}{d\,x^2}=-(m+n).$$

In such use y being the quantity of which the logarithm is taken, must be positive; hence the sign of $\frac{d^2\,u}{d\,x^2}$ will agree with the sign of $\frac{d^2\,y}{d\,x^2}$ in general, y being a common divisor in all dif. coefs. of u, or a common factor in all dif. coefs. of y.

In the particular case, u and y are maxima in necessary concurrence.

206. It is required to develop $y=\log.\ (x+h)$ by Taylor's Theorem, according to the powers of h, the expression log. being general, $\frac{1}{c}$ being the modulus.

$$\frac{d\,y}{d\,x}=\frac{1}{x}\cdot\frac{1}{c},$$

$$\frac{d^2\,y}{d\,x^2}=-\frac{1}{x^2}\cdot\frac{1}{c},$$

$$\frac{d^3\,y}{d\,x^3}=\frac{2}{x^3}\cdot\frac{1}{c},\ \text{etc.}$$

$$\therefore \log.(x+h)=\log.x+\frac{1}{c}\left(\frac{h}{x}-\frac{h^2}{2\,x^2}+\frac{h^3}{3\,x^3}-\frac{h^4}{4\,x^4}+,\ \text{etc.}\right)$$

SECTION XXVI.

EXAMPLES OF THE DIFFERENTIATION AND ANALYSIS OF EXPONENTIAL FUNCTIONS; INCLUDING EXAMPLES FROM COMPOUND INTEREST, AND INCREASE OF POPULATION.

1. From $a = b^x$ to find x, where a and b are any numbers, or numerical quantities:

$$\log. a = x \log. b \therefore x = \frac{\log. a}{\log. b} = \frac{\text{L}. a}{\text{L}. b}.$$

2. From $a^y = b^x$ to find $\frac{dy}{dx}$.

$$y \log. a = x \log. b \therefore \frac{dy}{dx} = \frac{\log. b}{\log. a}.$$

3. From $a = b^{mx}$ to find x.

$$\log. a = m x \log. b \therefore x = \frac{\log. a}{m \log. b}.$$

4. From $a = b^{-x}$ to find x.

$$\log. a = -x \log. b \therefore x = -\frac{\log. a}{\log. b}.$$

From this result we may infer that either a or b must be less than 1, or x must be held to be of a value opposite to that expressed by the sign in the function (Art. **97**).

5. From $a = b^x$ to find $\log. x$.

$$\log. a = x \log. b \therefore \log. x = \log.^2 a - \log.^2 b.$$

6. From $a = \frac{b^x}{p^{mx}}$ to find x.

$$\log. a = x \log. b - m x \log. p,$$

$$\therefore x = \frac{\log. a}{\log. b - m \log. p}.$$

7. From $y = \frac{a^x}{b^x}$ to find $\frac{dy}{dx}$.

$$\log. y = x \log. a - x \log. b,$$

$$d \log. y = \frac{dy}{y} = (\log. a - \log. b)\, dx,$$

$$\therefore \frac{dy}{dx} = \frac{a^x}{b^x} (\log. a - \log b).$$

8. From $y = a^x \log. x$ to find $\frac{dy}{dx}$.

Let $z = \log. x$, then $\log. y = x \log. a + \log. z$,

$$\therefore d \log. y = \frac{dy}{y} = dx \log. a + \frac{dz}{z};$$

now
$$\frac{dz}{z} = \frac{\frac{dx}{x}}{\log. x} = \frac{dx}{x \log. x},$$

$$\therefore \frac{dy}{dx} = \left(\log. a + \frac{1}{x \log. x}\right) a^x \log. x.$$

9. From $a = b^{\log. x}$ to find x.

$$\log. a = \log. x \log. b,$$

$$\therefore \log. x = \frac{\log. a}{\log. b} = \frac{L. a}{L. b};$$

x is therefore the natural number of which this quotient, $\frac{L. a}{L. b}$ is the logarithm.

10. From $y = a^x b^{nx}$ to find $\frac{dy}{dx}$.

$$\log. y = x \log. a + n x \log. b,$$

$$d \log. y = \frac{dy}{y} = \log. a \times dx + n \log. b \times dx,$$

$$\therefore \frac{dy}{dx} = a^x b^{nx} (\log. a + n \log. b).$$

11. From $ax = b^{xy}$ to find $\frac{dy}{dx}$.

Ans. $\frac{dy}{dx} = \frac{1 - \log. a + \log. x}{x^2 \log. b}$.

12. From $y = x^x$ to find $\frac{dy}{dx}$.

$$\log. y = x \log. x,$$

$$\frac{dy}{y} = \log. x \, dx + dx,$$

$$\therefore \frac{dy}{dx} = y \log. x + y,$$

$$= x^x (\log. x + 1).$$

13. Required the value of $\frac{a^x - b^x}{x}$ when $x = 0$.

(Art. 171.) $p' = \log. a \,.\, a^x - \log. b \,.\, b^x$.

$$q' = 1,$$

$$\therefore \frac{p'}{q'} = \log. a - \log. b = \log. \frac{a}{b}, \text{ Ans.}$$

14. Required the value of $\frac{a^x - x^a}{\log. a - \log. x}$ when $x = a$.

Ans. $(1 - \log. a) a^{a+1}$.

207. The successive terms of a series called *progression by quotient*, but as truthfully *progression by factor*, differ in their expression by a varying index only.

If a be the first term of such series, and q the factor, which, by association with a produces the second, q^2, q^3, etc., the factors which produce the succeeding terms, we have, after putting $q^0 = 1$ with a for the first term,

$$a q^0, \; a q, \; a q^2, \; a q^3, \; a q^4, \text{ etc.},$$

as an instance of the successive terms of such series. This may be called the general form. Any one of these terms is determined by a the *first term*, q the common ratio and n its index.

The compound interest of a sum of money for a term of time, the interest being supposed to be added to the principal at the end of each year, is represented by a term in such a series; and all its terms are the amounts as they are constituted at the end of each year.

For the better understanding of this, let us resolve q into $1 + \frac{r}{100}$, where 1 represents 1 dollar put at compound interest, and r the rate per cent.; consequently $\frac{r}{100}$ represents its interest for one year in the proper fraction of a dollar, and $1 + \frac{r}{100}$ constitutes the amount of the principal 1 dollar and its interest at the end of the first year. The quantity, a, is any sum of dollars, and is constant through all the terms; for all dollars at compound interest severally are like the 1 dollar mentioned as principal. The amount of the 1 dollar for the second year is evidently $\left(1 + \frac{r}{100}\right)^2$, for the third, etc., $\left(1 + \frac{r}{100}\right)^3$, etc., and for any sum a, the preceding expressions become $a\left(1 + \frac{r}{100}\right)^2$, $a\left(1 + \frac{r}{100}\right)^3$, etc.

In the succeeding context, we will use the small Roman r instead of the small italic r; as,

$$\mathrm{r} = \frac{r}{100};$$

but must return to the Italic r when we mention rate per cent., for then, in a distinct sense we make an integer of each unit of r.

We have, then, for successive amounts of the compound interest on any sum of dollars, a, at the end of successive years,

$$a(1+\mathrm{r}),\ a(1+\mathrm{r})^2,\ a(1+\mathrm{r})^3, \text{ etc.};$$

the sum put at interest, however, for the first year is entitled to the expression

$$a(1+\mathrm{r})^0 = a,$$

so that, as a series which may *ever need to be summed*, the term having the index n, is the $(n+1)$ st term.

If we call A the amount of principal and interest of a sum at compound interest for n years, which amount is nothing more than just that term of the series of which n is the index, we have the formulas:

$$A = a(1+\mathrm{r})^n, \tag{1.}$$

$$\therefore a = \frac{A}{a(1+\mathrm{r})^n}, \tag{2.}$$

$$n = \frac{L.A - L.a}{L.(1+\mathrm{r})}, \tag{3.}$$

$$\mathrm{r} = \left(\frac{A}{a}\right)^{\frac{1}{n}} - 1, \tag{4.}$$

$$S = \frac{a(q^n - 1)}{q-1} = \frac{a(q^{n-1}q - 1)}{q-1}, \tag{5.}$$

where S represents the sum of n terms of the series, $a\,q^0$, $a\,q$, $a\,q^2$, etc.; for the demonstration of formula (5.), we must refer to a complete treatise of algebra, since it is not like (2.), (3.), and (4.) deduced from (1.)

We are furnished now with the means of resolving questions which make n to be x, or a variable, and any of the other quantities to be y.

15. If 342 (a) dollars be put at compound interest at 5, (r) per cent., required how the amount (A) is increasing, compared with the years, at the end of $3\frac{1}{2}$ years.

From $$y = a\,(1 + \mathrm{r})^x$$

$$\therefore \frac{d\,y}{d\,x} = a \log.\ (1 + \mathrm{r})\ (1 + \mathrm{r})^{3\frac{1}{2}}$$

$$\frac{d\,y}{d\,x} = 342 \text{ Log. } \left(\frac{105}{100}\right) \left(\frac{105}{100}\right)^{3\frac{1}{2}} \times 2{,}30258 = 20, \text{ Ans.}$$

Hence it is increasing 20 times as fast as the years.

16. The sum of 1200 (a) dollars was put at compound interest till the amount (A) accrued to be 2525,82 dollars. If we first assume the number of years to have been 11, and immediately proceed to consider them more, how is the implied rate per cent. disposed to change, in accordance with the assumed variation of time?

On examining, in formula (4.), $\frac{A}{a}$, we observe that we have to do only with the value of this fraction, which, in the present problem, is 2,1048. We must remember the denominator of r is 100.

$$\therefore r = 100 \times (2.1048)^{\frac{1}{n}} - 100,$$

or, $$y = 100 \times (2.1048)^{\frac{1}{x}} - 100.$$

$$\therefore \frac{dy}{dx} = 100 \times 2.30258 \times \text{L.}\,2.1048 \times (2.1048)^{\frac{1}{x}} \times \left(-\frac{1}{x^2}\right)$$

$$= \frac{230{,}258 \times 0.32322 \times 1{,}07}{-121} = -{,}6583, \text{ Ans.}$$

Here we are obliged to remember the differential of the variable exponent.

Hence the rate per cent., which happens to be 7, is disposed to diminish 6583 ten thousandths of one per cent. The factor, 2.30258, will be remembered as the reciprocal of the modulus.

17. On the 1st day of January, 1864, the amount of principal and interest of a sum of money having been at compound interest, at 7 (r) per cent., was found to be 3579 (a) dollars; required to find the number of years distant before or after that date, when the compound interest of the sum, whatever it may have been, that was originally put at compound interest, should be found increasing 60 times as fast as the years.

With reference to the date given, the *amount* 3579 dollars, is not the A for the proposed investigation, but is a, the sum considered to be put at interest with reference to both future, and, as it were, past time. If n in the formula should be negative, then A as less than a, may be calculated for any past time. Now the question does not ask for the value of this A, but for its rate of change, which was just the variation of the interest only. Therefore, calling this A, or the interest either, y, we have

$$y = a\,(1 + \text{r})^x = 3579 \left(\frac{107}{100}\right)^x$$

$$\therefore \frac{dy}{dx} = 60 = 3579 \times 2.30258 \times \text{L.}\left(\frac{107}{100}\right)\left(\frac{107}{100}\right)^x,$$

$$\therefore x = \frac{\text{L. } 60 - \text{L. } 3579 - \text{L. }(\text{L. } 107 - \text{L. } 100) - \text{L. } 2.30258}{\text{L. } 107 - \text{L. } 100},$$

Now, L. 107 = 2.02938; L. 100 = 2 ∴ L. 107 — L. 100 = .02938, ∴ — L. (L. 107 — L. 100) = — L. .02938;

L.	60 =	+ 1, + 77815
— L.	3579 =	— 3, — 55376
— L.	.02938 =	+ 2, — 46805
— L.	2.30258 =	— 0, — 36222
		— 4, — 38403
		+ 3, + 77815
		— 60578

Now, —,60578 ÷ ,02938 = — 20,62 = — (20 years, 7 months, 13 days); and the date desired is May 18th, 1843, Ans.

18. (*a.*) If a body be put in motion through a resisting medium, by a force which impels it 10 rods in the first second of time, 9 rods in the next second, and so on, so that in any second of time the distance impelled shall be $\frac{9}{10}$ths of the distance in the preceding second, required how far it will go in all time.

With reference to formula (5.), S is the distance required, a is 10 rods, and q is $\frac{9}{10}$; now q being less than 1, its infinite power is 0. So that

$$S = -10 \div -\tfrac{1}{10} = 100 \text{ rods, Ans.}$$

This determines a limit for S; for the function $\frac{a\,q^x - a}{q-1}$, can have no mathematical maximum, as by definition.

(*b.*) Required how fast the body is moving at the end of 16 seconds; i. e., its *constant* rate, as it were, for an infinitesimal space of time, but mentioned in the language of *rate per second.*

Since S, the whole distance attained at the end of 16 seconds, will be varying just as the velocity, we will substitute y as velocity for S, for the purpose of differentiation, 16 or the number of seconds being x; then,

$$y = \frac{10\left[\left(\frac{9}{10}\right)^x - 1\right]}{-\frac{1}{10}} = -100\left(\frac{9}{10}\right)^x - 100$$

$$\frac{dy}{dx} = -100 \times 2{,}30258 \times \text{L.}\left(\frac{9}{10}\right)\left(\frac{9}{10}\right)^x$$

$$= 230{,}258 \times {,}04576 \times {,}1853$$

$$= 1{,}9526 \text{ rods, Ans.}$$

(c.) Required the actual distance moved through in the 17th second.

19. (a.) A man's property, on January 1, 1850, consisted of an investment of 1500 (a) dollars in stocks, paying an annual interest of 7 (r) per cent., but which is to remain invested. The remainder of his property was unemployed, with reference to producing any income, but was salable at the date mentioned for 2700 (a') dollars, and was destined to depreciate at the rate of 4 (r') per cent. annually, as indeed it had been previously. Required the date of the least value of the general balance of his property.

Let $y =$ the sum required, $= A + A'$, and $x =$ the number of years' difference of date. Then by the formula,

$$y = a\,(1 + \text{r})^x + a'\,(1 - \text{r}')^x$$

$$\therefore \frac{dy}{dx} = a\,(1 + \text{r})^x \log.(1 + \text{r}) + a'\,(1 - \text{r}')^x \log.(1 - \text{r}'),$$

$\therefore$ when y is a minimum, for we know from logical considerations that y has only a minimum, we have :

$$\left(\frac{1+\mathrm{r}}{1-\mathrm{r}'}\right)^x = \frac{-a' \log.(1-\mathrm{r}')}{a \log.(1+\mathrm{r})},$$

$$= \frac{-a' \mathrm{L}.(1-\mathrm{r}')}{a \mathrm{L}.(1+\mathrm{r})},$$

$$\therefore x = \frac{\mathrm{L}.a' - (\mathrm{Log.})^2 (1-\mathrm{r}') - \mathrm{L}.\,a - (\mathrm{Log.})^2 (1+\mathrm{r})}{\mathrm{L}.(1+\mathrm{r}) - \mathrm{L}.(1-\mathrm{r}')};$$

where $\quad -(\mathrm{Log.})^2 (1-\mathrm{r}') = \mathrm{L}.\left(-\mathrm{L}.(1-\mathrm{r}')\right)$,

and is therefore a positive quantity, L. (1 — r′) being negative, and — L. (1 — r′) being positive, and in real arithmetical expression, being without the — sign.

The advantage of electing to place the negative sign before L. (1 — r′), in preference to any other factor, is manifest.

(*b*.) Required the date at which the two species of property become of equal values; and also that value.

20. (*a*.) A certain country consists of two districts, Eastern and Western. On a certain date the Eastern contained 6,272,000 inhabitants, who were, and had been increasing at the rate of 16 per cent. in 10 years. The Western contained at the same date 9,035,000 inhabitants, who were and had been decreasing at the rate of 5 per cent. in 8 years. Required the different date before or after, of the minimum population of the country.

(*b*.) Required the date of like population of the districts.

(*c*.) Required the date at which the Eastern district must contain nine times as many inhabitants as the Western.

21. Required the number by which, if we divide *a* and raise the quotient to the power indicated by that divisor, the power shall be a maximum.

Let $\qquad y =$ the power;

then $\qquad y = \left(\frac{a}{x}\right)^x;$

$$. \log. y = x \log. \frac{a}{x} = x \log. a - x \log. x,$$

$$\therefore \frac{d\,y}{d\,x} = y \log. a - y - y \log. x,$$

$$\therefore \log. \frac{a}{x} = 1 \text{ when } \frac{d\,y}{d\,x} = 0,$$

$$\therefore \text{L.}\, \frac{a}{x} = .43429,$$

$$\therefore \text{L.}\, x = \text{L.}\, a - .43429,$$

$$\text{L.}\, x = \text{L.}\, a - \text{L.}\, 2.71828,$$

$$\therefore x = \frac{a}{2.71828}.$$

208. In the common arithmetical computation of compound interest for cases when there are months and days, additional to entire years as the time, the usual direction is to find first the amount of principal and interest for the entire years, on which, as principal, to compute the interest for such additional months and days. But this course will always give the entire compound interest somewhat too great, because it assumes that the interest is to accumulate uniformly during such months and days by a rate that is directly proportional to the result that would accrue for an additional year.

When a sum is put at simple interest for one year at a given rate per cent., the virtual amount of principal and interest is greater, relatively to the time, during the later or last months, than during the first and earlier months, for the reason that the unpaid interest of the early month is itself on interest during the later month. The idea of calling the use of money worth a given per cent. for a year, is therefore a compromise for ready convenience.

Let us now actually compute the true interest of 100 dollars for the *successive months* of a year, at six per cent. per year by reducing n in the formula (1.) p. 197 to months, which is done by giving to it the numerator 12. Formula 4th then becomes, when r is $\frac{6}{100}$, so far as the constitution of A is concerned, or the amount for one year, but still remains as r, for the *first month's rate*,

$$\mathrm{r} = \left(\frac{106}{100}\right)^{\frac{n}{12}} - 1 = \left(\frac{A}{a}\right)^{\frac{n}{12}} - 1,$$

whence, when $n = 1$, we have

$$\mathrm{r} = {,}00487\text{; and } r = {,}487\text{;}$$

a decimal fraction, of which the unit would be 1 per cent.; or it may be read 48 cents 7 mills, as the interest of 100 dollars for 1 month, at *nominal* six per cent., and indeed *virtual* six per cent. for 1 year by the language of compromise. If we were to speak of a mathematical per cent., then multiplying ,487 by 12, we arrive at $5\frac{844}{1000}$ as the mathematical per cent. for the rate of interest of money for the first month, when at "interest at six per cent. the year."

Calling $n = 2$, and obtaining another result for r, and subtracting that obtained, we have for the interest of 100 dollars for the second month, 49 cents. Continuing in this manner, we find, as the result, the interest for the twelve successive months to be, in cents, 48.7; 49.0; 49.2; 49.4; 49.6; 49.8; 50.1; 50.3; 50.5; 50.8; 51.1; 51.4, which, added, make 600 cents, or 6 dollars.

If n be considered to have values, then, between integral numbers for years, or to be a quantity having "flowing" values, we are enabled to derive true computations for compound interest for other lengths of time than entire years. In this consists the value of that formula, and of this analysis afforded by the differentiation of exponentials.

22. Required the true time in which a sum of money becomes doubled when put at compound interest, at 5 per cent., and how much more it is than by the arithmetical way.

Ans. 14 yrs. 2 mo. 15 da., being 2 da. more.

23. The white population of the United States, from June 1, 1830, to June 1, 1840, increased 34 per cent. Required its annual ratio of increase, and in what time it must have become doubled.

Ans. $2\frac{966}{1000}$ per cent. Doubled in 23 yrs. 8 mo. 12 days.

24. Required the true compound interest of 360 dollars for 5 years, 6 months, and 24 days, and how much less the true is than the usual arithmetical result.

Ans. $137.80. The difference, 34 cents less.

A true but impracticable definition of compound interest, ignoring the termination of entire years, as essential, may be: A sum of money is said to be put at compound interest for a period of time, when the value of the use of a unit of it for a time however short, but definitely stated, is agreed upon at a rate (or proportion of that unit), and such rate or interest is added to that unit at the end of such period, and the use of the sum of them for a similar period of time is estimated at the same rate; and this second interest is added to the sum mentioned as the sum for use during a third similar period of time, and so on to the end of the last similar period of time, and the formation of a final sum, called amount of principal and the interest.

SECTION XXVII.

DIFFERENTIATION OF CIRCULAR FUNCTIONS.

[Unlike all the previous sections, the present presupposės the principles of Analytical Trigonometry to be understood.]

209. A *circular independent variable* may be an arc of a circle, or its sine, tangent, or other trigonometric line referred to the arc.

A *circular function* is a function of some linear trigonometric variable; it may be the arc itself, when the arc is not assumed as independent variable; it may be a sine, tangent, etc., of the variable arc.

210. Trigonometrical quantities are all to be considered numerical linear amounts in their result. Quantities strictly algebraic, as factors, etc., may contribute to this result; as, m in sin. $m\,x$, n in n tan. x. These quantities, when they are powers, and when their amounts agree with certain areas, are nevertheless to be regarded as linear amounts, or multiplications of a line. Hence trigonometrical quantities are special in kind, and not general, like arithmetical or algebraic quantities.

When an arc is made variable, we may call it the arc x. If the radius of the circle be 1, or not, its variable arc is of unlimited length, by repetitions of itself if need be. All the principles of trigonometry, and of the differentiation of circular functions, are intended to apply to this unlimited arc.

The expressions for circular functions, sin. x, tan. x, etc., are intended to signify the sine, tangent, etc., of the arc which is, in length, x of the units of which the radius is 1, unless otherwise expressed. We can hardly call x sin. a a circular variable function, the variable x, not being restricted to the circle.

It is necessary to make the interpretation of the whole expression intended as the circular function, however the variable x may occur in it, contribute to a homogeneous result. In x sin. x, for instance, the prefixed x, being a factor to sin. x, must be abstract numerical, but of the same numerical value as that of x in sin. x, so that the result is a linear amount. In $x +$ sin. x it is necessary to interpret the isolated x as the arc to which another linear amount, sin. x, is added.

Powers of the sine of the arc x, cosine of the arc x, etc., are expressed by the exponent attached to the prefix, as, $\sin.^2 x$, $\text{cosec.}^n x$. This leaves a distinctive signification for sin. x^2, etc., which is obvious, and requires no use of parentheses.

In the expression, sin. x, it is scarcely necessary to remark that the prefix sin. is not mathematically separable from x, and we will not adopt that inverse notation which from $y = \sin.^{-1} x$ would attempt to derive sin. $y = x$. Indeed, we are already committed to regarding $\sin.^{-1} x$ as equivalent to $\frac{1}{\sin. x}$.

In the function $x^{\cos. x}$ the exponent must be taken in its numerical sense, apart from being linear, and the root x must be the same arc of which the cosine is intended, raised in its numerical sense to the power cos. x; the resulting power may at last be taken as that of the arc.

The expression log. sin. x must be abstract numerical.

The expression sin. log. x must be linear, and the arc intended must be of the numerical length, log. x, radius being 1.

211. It is evident that the differential of an arc of the length zero, may be called identical with the differential of the tangent of it, or chord of it.

We are to understand sin. (sin. x) to signify the sine of the arc which is of the length sin. x, which is of course a less arc than x of the same circle. As with logarithmic functions we use $(\log.)^2 x$ for log. (log. x), so we will use $(\sin.)^2$ for sin. (sin. x). The second power of sin. x will be $\sin.^2 x$ without the parenthesis.

Such an expression as $a^{\sin. x}$ may be called an exponential circular function.

Circular, Logarithmic, and Exponential Functions, are called Transcendental Functions.

212. In order to differentiate sin. x, we have for radius 1, if a be any arc, and b be any additional arc, by the ratio of corresponding parts of similar right-angled plane triangles:

$$\sin. (a + b) - \sin. a : \tan. b :: \cos. a : 1;$$

that is,

$$\sin. (x + h) . - \sin. x : \tan. h :: \cos. x : 1;$$

but if $h = 0$, sin. $(x + h)$ — sin. x becomes d sin. x, and tan. h becomes $d\,x$, or differential of the arc x.

$$\therefore d \sin. x : d\,x :: \cos. x : 1$$

$$\therefore d \sin. x = \cos. x\, d\,x.$$

213. If the arc be designated otherwise than by x, as for instance by x^n, or $x + x^3$, etc., then, instead of $d\,x$, we must substitute the differential of that algebraic or other function of x, which does designate the intended arc.

214. If, moreover, the function to be differentiated be some function of sin. x, as sin.$^2\, x$, an instance of which will immediately follow, we must differentiate it as any algebraic power, and make *the differential of the root* a factor in the differential required.

215. In order to differentiate cos. x we have

$$\cos. x = (1 - \sin.^2 x)^{\frac{1}{2}},$$

$$\therefore d \cos. x = (d\,(1 - \sin.^2 x)^{\frac{1}{2}}),$$

$$= -\tfrac{1}{2}\,(1 - \sin.^2 x)^{-\frac{1}{2}} \times 2 \sin. x \cos. x\, d\, x,$$

$$= -\frac{\sin. x \cos. x}{(1 - \sin.^2 x)^{\frac{1}{2}}},$$

$$= -\frac{\sin. x \cos. x}{\cos. x}\, d\, x = -\sin. x\, d\, x.$$

216. In order to differentiate tan. x we have

$$d \tan. x = d\,\frac{\sin. x}{\cos. x} = \frac{\cos. x\, d \sin. x - \sin. x\, d \cos. x}{\cos.^2 x}$$

$$\therefore d \tan. x = \frac{\cos.^2 x + \sin.^2 x}{\cos.^2 x}\, d\, x\,;$$

but $$\cos.^2 x + \sin.^2 x = 1$$

$$\therefore d \tan. x = \frac{1}{\cos.^2 x}\, d\, x = \sec.^2 x\, d\, x.$$

217. In order to differentiate cot. x we have

$$d \cot. x = d\,\frac{1}{\tan. x} = -\frac{d \tan. x}{\tan.^2 x}$$

$$= -\frac{\sec.^2 x}{\tan.^2 x}\, d\, x = -\operatorname{cosec.}^2 x\, d\, x.$$

218. In order to differentiate sec. x we have

$$d \text{ sec. } x = d \frac{1}{\text{cos. } x} = \frac{\text{sin. } x}{\text{cos.}^2 x} d x$$

$$= \frac{\text{tan. } x}{\text{cos. } x} d x = \text{tan. } x \text{ sec. } x \, d x.$$

219. In order to differentiate cosec. x we have

$$d \text{ cosec. } x = d \frac{1}{\text{sin. } x} = - \frac{\text{cos. } x}{\text{sin.}^2 x} d x$$

$$= - \frac{\text{cos. } x}{\text{sin. } x} \times \frac{1}{\text{sin.} x} = - \text{cot. } x \text{ cosec. } x \, d x.$$

220. Therefore, by recapitulation:

$$\begin{aligned}
d \text{ sin. } x &= \text{cos. } x \, d x \\
d \text{ cos. } x &= - \text{sin. } x \, d x \\
d \text{ tan. } x &= \text{sec.}^2 x \, d x \\
d \text{ cot. } x &= - \text{cosec.}^2 x \, d x \\
d \text{ sec. } x &= \text{tan. } x \text{ sec. } x \, d x \\
d \text{ cosec. } x &= - \text{cot. } x \text{ cosec. } x \, d x.
\end{aligned}$$

If each of these functions be y, the expressions for their first dif. coefs. are obvious.

221. Whenever occasion may require the differentiation of a circular function, for radius R other than 1, it is necessary to employ R in the place of 1 in the course of the method of determining the differential, because, although $1 = 1^2$, this would not be true of R and R^2.

1. Required to develop sin. x by Maclaurin's Theorem:

$$y = \text{sin. } x \; . \; . \; . \; . \; . \; . \quad (y) = 0$$

$$\frac{d y}{d x} = \text{cos. } x. \; . \; . \; . \; . \; . \quad \left(\frac{d y}{d x}\right) = 1$$

$$\frac{d^2 y}{d x^2} = -\sin. x \ldots\ldots \left(\frac{d^2 y}{d x^2}\right) = 0$$

$$\frac{d^3 y}{d x^3} = -\cos. x \ldots\ldots \left(\frac{d^3 y}{d x^3}\right) = -1$$

$$\frac{d^4 y}{d x^4} = \sin. x \ldots\ldots \left(\frac{d^4 y}{d x^4}\right) = 0$$

$$\frac{d^5 y}{d x^4} = \cos. x \ldots\ldots \left(\frac{d^5 y}{d x^5}\right) = 1, \text{ etc.}$$

$$\therefore \sin. x = x - \frac{x^3}{1.2.3} + \frac{x^5}{1.2.3.4.5} - \frac{x^7}{1.2.3.4.5.6.7} +, \text{ etc.}$$

2. Required to develop cosine x.

$$y = \cos. x \ldots\ldots\ldots (y) = 1$$

$$\frac{d y}{d x} = -\sin. x \ldots\ldots \left(\frac{d y}{d x}\right) = 0$$

$$\frac{d^2 y}{d x^2} = -\cos. x \ldots\ldots \left(\frac{d^2 y}{d x^2}\right) = -1, \text{ etc.}$$

$$\therefore \cos. x = 1 - \frac{x^2}{1.2} + \frac{x^4}{1.2.3.3} - \frac{x^6}{1.2.3.4.5.6} +, \text{ etc.}$$

222. If we take the expression for the developed sin. x, and differentiate it, we have

$$d \sin. x = d x - \frac{x^2 d x}{1.2} + \frac{x^4 d x}{1.2.3.4} -, \text{ etc.,}$$

$$= (1 - \frac{x^2}{1.2} + \frac{x^4}{1.2.3.4} -, \text{ etc.}) \, d x$$

$= \cos. x \, d x$, as by development of cos. x.

223. In like manner if we take the expression for the developed cos. x, and differentiate it, we have

$$d \cos. x = -(x + \frac{x^3}{1.2.3} - \frac{x^5}{1.2.3.4.5} +, \text{ etc.,}) \, d x$$

$= -\sin. x \, d x$ as by development of sin. x.

The summation of the series expressive of sin. x and cos. x, for particular lengths of the arc x, must give the natural sine, natural cosine of such arc. The sine and cosine of one arc being obtained, the sine and cosine of m times such arc may be found by the following formulas, of which we omit the demonstration:

$$\sin. m x = m \cos.^{m-1} x \sin. x - \frac{m(m-1)(m-2)}{2.3} \cos.^{m-3} x$$

$$\sin.^3 x +, \text{etc.}$$

$$\cos. m x = \cos.^m x - \frac{m(m-1)}{2} \cos.^{m-2} x \sin.^2 x +$$

$$\frac{m(m-1)(m-2)(m-3)}{2.3.4} \cos.^{m-4} x \sin.^4 x -, \text{etc.}$$

224. In a manner similar to that of sin. x and cos. x, may tan. x, cot. x, etc., be developed. Such are developments of sin. x, cos. x, etc., depending on a portion x of the arc as assumed variable, useful when the arc is known. But we may equally develop the arc x, in terms of some function of it, sin. x, cos. x, tan. x, etc., and in doing so, while we will preserve the notation as already used, we are obliged to regard the function y as the independent variable, and x the arc as the dependent variable; whence $\frac{dx}{dy}$, in such case, becomes the reciprocal of those inferred from Art. 219 for $\frac{dy}{dx}$,

3. Required to develop x in $y = \tan. x$.

$$y = \tan. x \therefore \text{when } y = 0 \quad . \quad . \quad (x) = 0$$

$$\frac{dx}{dy} = \frac{1}{\sec.^2 x} = \frac{1}{1+y^2} \quad . \quad . \quad . \quad \left(\frac{dx}{dy}\right) = 1$$

$$\frac{d^2 x}{dy^2} = -\frac{2y}{(1+y^2)^2} \quad . \quad . \quad . \quad . \quad \left(\frac{d^2 x}{dy^2}\right) = 0$$

$$\frac{d^3 x}{d y^3} = -\frac{2}{(1+y^2)^2} + s\,y. \quad . \quad . \quad . \quad \left(\frac{d^3 x}{d y^3}\right) = -2$$

$$\frac{d^4 x}{d y^4} = \frac{2^3\, y}{(1+y^2)^3} + s'\,y \quad . \quad . \quad . \quad . \quad \left(\frac{d^4 x}{d y^4}\right) = 0$$

$$\frac{d^5 x}{d y^5} = \frac{2^3 . 3}{(1+y^2)^3} \pm s''y \quad . \quad . \quad . \quad . \quad \left(\frac{d^5 x}{d y^5}\right) = 2^3 . 3$$

where s, s', s'' are quantities factors to y,

$\therefore x = \tan. x - \frac{1}{3} \tan.^3 x + \frac{1}{5} \tan.^5 x - \frac{1}{7} \tan.^7 x +$, etc.

If now x be an arc of 45°, $\tan. x = 1 =$ radius,

$$\therefore \text{arc } 45^\circ = 1 - \tfrac{1}{3} + \tfrac{1}{5} - \tfrac{1}{7} + \tfrac{1}{9} -, \text{ etc.,}$$

which, from its slow convergency, is not readily summed. Its sum is, in terms of radius $= 1$, the length of the arc of 45°, or the eighth part of the circumference of the circle.

By the aid of the trigonometrical formula,

$$\tan. (a + b) = \frac{\tan. a + \tan. b}{1 - \tan. a \tan. b},$$

we may obtain Euler's series for the same purpose, which is much more convergent. For when $a + b = 45°$, tan. $(a + b) = 1$, therefore

$$\tan. a + \tan. b = 1 - \tan. a \tan. b.$$

If now either tan. a or tan. b were given, the other becomes determinable from this equation. Thus, if we suppose

$$\tan. a = \frac{1}{n}, \text{ then } \frac{1}{n} + \tan. b = 1 - \frac{\tan. b}{n},$$

$$1 + n \tan. b = n - \tan. b \therefore \tan. b = \frac{n-1}{n+1}$$

Developing respectively tan. $a = \frac{1}{n}$ and tan. $b = \frac{n-1}{n+1}$, by the method last found for tan. x we have

$$a = \frac{1}{n} - \frac{1}{3n^3} + \frac{1}{5n^5} - \frac{1}{7n^7} +, \text{ etc.}$$

$$b = \frac{n-1}{n+1} - \frac{n-1}{3(n+1)^3} + \frac{n-1}{5(n+1)^5} - \frac{n-1}{7(n+1)^7} +, \text{ etc.}$$

The value of n being arbitrary if we make $n = 2$, for this value makes the two series converge with a near equality, we have, if a sufficient number of terms be summed,

$$4(a+b) = 45° \times 4 = 3.141592653589793,$$

for the ratio of the semi-circumference of a circle to radius, or of the whole circumference to the diameter.

225. We have already given the development of the sine of an arc in terms of the arc. If it be desired to calculate numerically the natural sine of an arc designated by degrees, minutes, and seconds, as for instance for 27° 10′ 0″, it is necessary to translate this designation by degrees, etc., into numerical parts of radius 1. Thus, 27° 10′ = 1630′, and 180° = 10800′, and $\frac{163}{1080}$ of 3.1415926 is .47414777 the length of the arc of which the natural sine is required.

If now we select for use .4741 as the arc, and sum merely three terms of the development, we shall have the usual tabular amount to five decimal places:

$$\text{nat. sin. } 27° 10' = .4741 - \frac{.4741^3}{2.3} + \frac{.4741^5}{2.3.4.5} -, \text{ etc.}$$

$$= .45658$$

By logarithms we have

Log. .4741	$-1.+67587$
	3
Log. $.4741^3$	$-1.+02761$
— Log. $(2.3=6)$	$.-77815$
Log. .01776	$-2.+24946$

Again,

Log. .4741	$-1.+67587$
	5
Log. $.4751^5$	$-2.+37935$
— Log. $(2.3.4.5=120)$.	$-2.-07918$
Log. .0001996	$-4.+30017$

Now,

$$\begin{array}{r} .47414 \\ -\ .01776 \\ \hline .45638 \\ +\ .0001996 \\ \hline .45658 \end{array}$$ nat. sine required.

From the natural sine of an arc we easily obtain its natural cosine, natural tangent, cotangent, secant, and cosecant.

4. Required to differentiate $y = \sin.\ (a + m\,x)$.

Ans. $d\,y = m \cos.\ (a + m\,x)\ d\,x$.

5. Required the dif. coef. of $y = \sin.\ x \tan.\ n\,x$.

Ans. $\frac{d\,y}{d\,x} = \cos.\ x \tan.\ n\,x + n \sin.\ x \sec.^2 n\,x$.

6. Required dif. coef. of $y = \sin.^2 x$.

Ans. $\frac{d\,y}{d\,x} = 2 \sin.\ x \cos.\ x = \sin.\ 2\,x$.

7. Required dif. coef. of $y = \log. \sin. x$.

Ans. $\frac{dy}{dx} = \frac{\cos. x}{\sin. x} = \frac{1}{\tan. x}$.

8. From $\sin. x = \cot. y$, to find $\frac{dy}{dx}$.

$$\cos. x\, dx = - \operatorname{cosec.}^2 y\, dy$$

$$\therefore \frac{dy}{dx} = - \frac{\cos. x}{\operatorname{cosec.}^2 y} = - \cos. x \sin.^2 y.$$

9. Required the value of $\frac{\sin. x}{\tan. x}$ when $x = 0$.

Ans. 1.

10. Required value of $\frac{\tan. x - \sin. x}{\sin. x^3}$ when $x = 0$.

Ans. $\frac{1}{2}$.

11. Required the values of x when $\sin. x$ is a maximum and minimum.

Ans. x is a maximum at 90°, 450°, etc., and a minimum at 270°, or — 90°, 630°, or — 450°, etc.

12. Required the value of x, when $y = \sin. x - \sin.^2 x$ is a maximum.

$$\frac{dy}{x} = \cos. x - 2 \sin. x \cos. x = 0,$$

$$\therefore 1 - 2 \sin. x = 0,$$

$$\therefore 1 = 2 \sin. x,$$

$$\therefore x = \text{arc. } 30°.$$

13. From $y = x^{\cos. x}$ to find $\frac{dy}{dx}$.

$$\log. y = \cos. x \log. x,$$

$$\frac{dy}{y} = \frac{\cos. x\, dx}{x} - \sin. x \log. x\, dx,$$

$$\therefore \frac{d\,y}{d\,x} = x^{\cos. x - 1} \cos. x - x^{\cos. x} \sin. x \log. x.$$

14. From $\cot. y = x^{\cos. x}$ to find $\frac{d\,y}{d\,x}$.

$$\text{Ans.} \quad \frac{d\,y}{d\,x} = \frac{(x \sin. x \log. x - \cos. x) \cot. y}{x \operatorname{cosec.}^2 y}.$$

SECTION XXVIII.

GEOMETRICAL ILLUSTRATIONS OF THE VALUES OF FUNCTIONS, AND THE CORRESPONDING VALUES OF THEIR VARIABLES; ALSO OF THE VALUES OF DIFFERENTIAL COEFFICIENTS, MAXIMA AND MINIMA, ETC.

We have already passed in review the elementary principles of the differential calculus to a liberal and comprehensive extent. But we have purposely deferred geometrical illustration, because, if we had hitherto suffered it to engross attention, it might have become an evil so great as to require decisive counteraction. The illusion is apt to prevail that the differential calculus relates only to lines, or forms; the geometrical construction has a parallel and independent nature.

226. Every algebraic quantity or expression not imaginary, consisting of an aggregate of terms, may, by performing the algebraic indications, be considered resolved into, or constructed as, one resultant numerical amount; first as abstract units, inclusive also of fractional expressions of units, next as units of length, such as may be represented severally as each a straight line, or collectively as a continuous straight line. Thus, the numerical units, understood to be intended by the algebraic quantity a, may be represented by a straight line. But it is equally true of the product $a\,b$, of the quotient $\frac{a}{b}$, or indeed of such an aggregate as

$$\frac{a^2}{b} + \sqrt{c\,a} - b^3, \text{ etc.}$$

Methods, however, are pointed out in analytical geometry, by which the values of certain expressions may be illustrated geometrically, and eliminating all considerations about irrational values, by the methods adopted for the construction.

Next, let the expression contain, besides constants, the variable x, that is, be a function of one variable x, and it becomes evident that we can determine a straight line of definite length as its value, by supposing a value for x. Let the straight line, P P′ (Fig. 1), represent one of these values, and let it have the more perpendicular position for a conventional reason, its universal adoption by geometers to represent an ordinate of a plane curve, or of such line as may be determined by ordinates. This leaves a chance for some different line, I P, conventionally adopted as more horizontal to be the record of the value of x, and they have the point of meeting P in common. The position of the line P P′, with regard to the angle it makes with I P, may be any; it may be perpendicular to I P, but need not be necessarily so. All these lines and points are to be supposed to be in one plane.

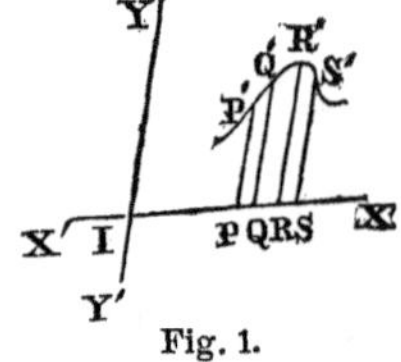

Fig. 1.

Next, let some other value be assumed for x, and for simplicity, a value greater by a small amount = P Q, so that we now have $x =$ I Q, and let the corresponding value of Fx be deduced, which may be Q Q′, and let it be parallel to P P′, and the point Q be taken in I P produced indefinitely toward X. We should adopt the line I X as dividing positive values of Fx from negative values of it. Let those values of Fx on the side of I X toward P′, Q′ be positive, and negative values of Fx will be found on the other side of I X. Let still a new value I R be assumed for x, and a corresponding value of Fx be found and be represented by R R′. Indeed, let such a number of values of x be assumed, and the corresponding values of Fx be found and located, sufficiently near together, as to give a complete illustration of the nature of these successive values.

In order to accommodate such conditions as grow out of x at negative values, I being the origin of values, and positive being conventionally toward the right, X, we need the line I X extended indefinitely to X′.

Let us also, through I at any angle with I X, draw the straight line Y Y′ as an original line of indefinite extension, parallel to which we will suppose we have drawn P P′, Q Q′, R R′, etc.

The intersecting lines X X′ and Y Y′, called by these designations as suggestive of the values of x measured on or parallel with the former, and of the function y, measured on or parallel with the latter, are the same as the axes of coördinates in analytical geometry. We may call them lines of reference, which are to be supposed existing with reference to all the constructions of the values of functions of one independent variable we may wish to make in this section.

Now, the continuous line which shall join the points P′, Q′, R′, and all other necessary points determined in the same way, is the particular line sought, since every point in it, when referred to Y Y′ by a straight line parallel with X X′, and to X X′ by a straight line parallel with Y Y′, shows a value of the function and of the variable in correspondence. This line may be called the *locus* of the values of the variables, or *locus* of the equation.

The positions of the four conditions alluded to in Art. **97**, become quite evident in the construction.

Now, the tediousness of this way of proceeding is very much relieved by the adoption of certain principles depending on the character of the function, and on the availability of differentiation.

227. The construction of the values of a function of a variable derivable from or referable to the general equation of the First Degree, viz.,

$$Ax + By + C = 0,$$

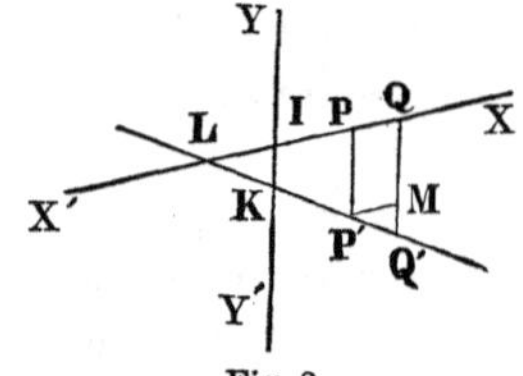

Fig. 2.

is simplest by making $x = 0$, when we have $y = -\frac{C}{B}$, which value set off on I Y′ determines K (Fig. 2); making $y = 0$, we derive $x = -\frac{C}{A}$, which determines the negative value I L, for x. A straight line through K and L, extended indefinitely, both ways, is the line sought.

If P Q, that is P′ M, be the increment h of the variable, when having the value I P′, Q′ M becomes the decrement of the function when passing the value P P′ (supposing P′ M drawn parallel with I X), that is,

$$\frac{F(x+h)}{h} = \frac{Q' M}{P' M},$$

and this quotient, in reference to the equation of the first degree, happens to be of the same value as

$$\frac{d\,F\,x}{d\,x}, \text{ or, } \frac{d\,y}{d\,x};$$

that is, as the *differential coefficient* of the function, because the value of this ratio does not change by making $h = 0$.

As examples, let each of the following explicit or implicit functions of x be constructed as specified.

1. $y = 8 + 2x$.
2. $5y - 7 + 4x = 0$.
3. $3(4\frac{1}{2} + 2x) + \frac{y}{6} = 0$.
4. $y + x = 0$.
5. $\frac{x}{y} + 6 = 0$.

228. We can now illustrate geometrically how two algebraic expressions, each containing a quantity x called unknown, when equated with each other as in the solution of some algebraic problems, render a determinate value for x, and exclude it from being a variable. We may construct independently each of these expressions as a function of x. The point of a common value of each of these functions determines the required value of x.

This principle is general with reference to equations of different degrees, and more values of x than one. At present, however, let the illustration be of an equation of the first degree.

Given the algebraic equation of virtually the first degree, viz.,

$$a x + b = a' x + b',$$

to construct the value of x, which satisfies the condition, without transposition of one member.

Let I M (Fig. 3) be the value b, that is, what the first member becomes when $x = 0$, and M M′ be the *locus* of all values of $a x + b$; let I L $= b'$, and L L′ be the *locus* of all values of $a' x + b'$. If they intersect, let P′ be the point of intersection; draw P P′ parallel to Y Y′, and we have I P, the value of x required. If L L′ and M M′ do not intersect, in which case a must be

$$a = a',$$

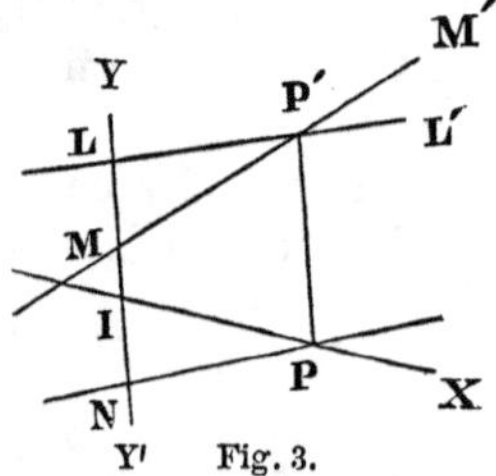

Fig. 3.

x will be indeterminate, L L′ M M′ either agreeing or being parallel to each other.

If, however, transpositions of one member of the equation first take place, the construction will be one straight line, which must be N P, I N being equal to I M — I L; that is, to $b - b'$ in

$$(a - a')\,x + (b - b') = 0;$$

considering $a > a'$ and $b' > b$.

229. In the arithmetical rule of Double Position, in which we operate without the possession of a visible written equation, we virtually have, in the conditions of questions offered for solution, a function (referable to an equation of the first degree), equal to zero, to find x, and we are directed to suppose any two numbers for the unknown quantity, and to test each of them in the conditions with reference to finding the result, zero; the variations from zero we call *errors;* whence by the use of the supposed numbers, and the errors, we derive the value of the unknown quantity.

In Fig. 4, let N P be construction of

$$A\,x + B = 0,$$

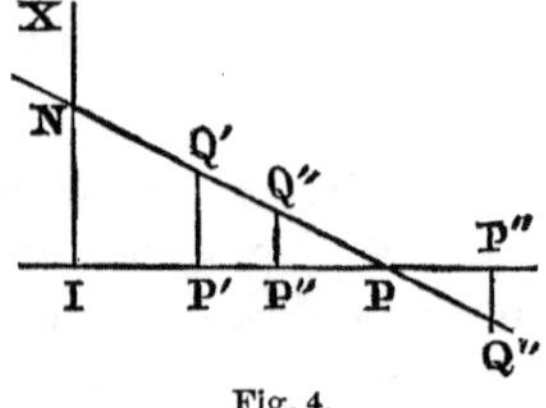

Fig. 4.

to find x or I P; we suppose I P′ = S, to be I P, and find the error P′ Q′ = E; next we will suppose I P$''$ = S$'$ to be I P, and find the error P$''$ Q$''$ = E$'$. Whence we have

$$B : x :: E \pm E' : S' \pm S.$$

This is an algebraical form of the rule, which, in arithmetical language, is necessarily stated with much circumlocution.

230. It is evidently impossible to draw a line representing the differential dx of a variable x, and another representing dy, the differential of the function y, each of such lines being zero in length; hence we cannot exhibit in visible amounts the ratio $\frac{dy}{dx}$. But we can exhibit linear amounts of which the value of their ratio is exactly the same as $\frac{dy}{dx}$. Since there is a presumption that the construction of the values of functions in general, of a single variable, may be by a line or lines not necessarily straight, but curved, although without regard to a special function we can determine nothing of its law, let a part of such line be P$''$ P$'$ S, Fig. 5, and let T T$'$ be a straight line meeting it at P$'$, and agreeing with it in the nearest vicinity of P$'$; then,

I P being x, P P$'$ being y, let P Q = P$'$ Q$'$ = h, and we have Q P$''$ = $F(x+h)$ and Q$'$ P$''$ = $F(x+h) - Fx$; hence we have

$$\frac{F(x+h) - Fx}{h} = \frac{P''Q}{P'Q'},$$

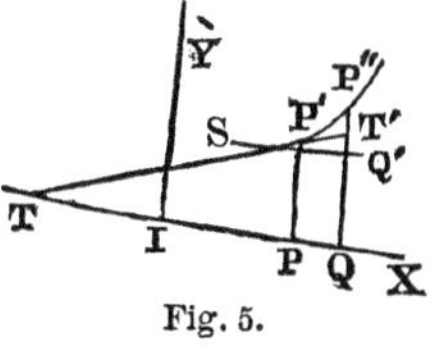

Fig. 5.

and when $h = 0$, although T$'$ Q$'$, P$'$Q$'$ each become 0, we have

$$\frac{dy}{dx} = \frac{T'Q'}{P'Q'} = \frac{PP'}{TP}.$$

In case $y =$ a maximum or minimum, we evidently have $TP = \infty$, TT' being parallel with IX, so that

$$\frac{dy}{dx} = 0.$$

Now, the second and succeeding dif. coefs. are collectively represented in value, by the ratio

$$\frac{P''T'}{P'Q'}$$

in the value it assumes when $P'Q' = h$, becomes zero.

With a view, however, to make originally the construction of a given function of a single variable, and to determine the direction of the line $P'Q'R'S'$ (Fig. 1) through any point P', the obvious direction is to determine the value of $\frac{dy}{dx}$ for that point, and by this value construct the course as a straight line for the immediate vicinity of the point, and through the point.

When the axes of reference IY and IX are rectangular, the value of $\frac{dy}{dx}$ for any point P' is the trigonometrical tangent of the angle, which the straight line touching the curve at that point makes with IX, called the axis of x in analytical geometry.

231. Before proceeding to the geometrical construction of equations of the second and higher degrees, we properly give attention to illustrations of maxima and minima of functions of one independent variable.

The character of a maximum is shown in Fig. 6, by observing that there is a value at P' greater than the nearest contiguous values on either side of it. The maximum value of the function is PP'.

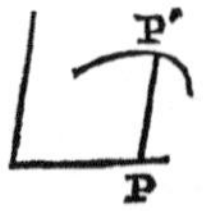

P'

P

Fig. 6.

A minimum is shown at Fig. 7. A maximum and minimum of the same function is shown in Fig. 8. In these cases $\frac{dy}{dx}$ is supposed equal to zero.

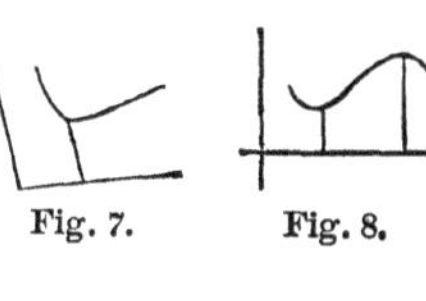

Fig. 7. Fig. 8.

In Fig. 9 is shown the character of a maximum, and in Fig. 10 of a minimum, when $\frac{dy}{dx} = \pm \infty$. These cases evidently come within the definitions.

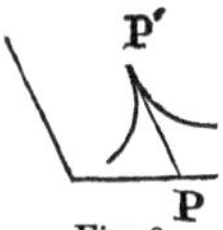

P′ P

Fig. 9. 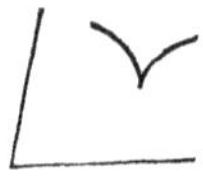 Fig. 10.

We show, in Figs. 11 and 12, cases illustrative of $\frac{dy}{dx} = 0$, while there is neither a maximum nor minimum. Instances by actual functions are,

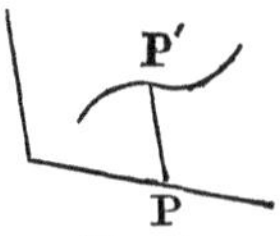

P′ P Y P′ X′ P I

Fig. 11. 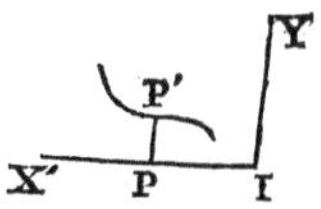 Fig. 12.

by Fig. 11, $y = a + (x - b)^3$,

by Fig. 12, $y = a - (x + b)^3$.

Another instance is given in Fig. 13, a corresponding function being

$$y = b + (x - a)^{\frac{3}{2}}.$$

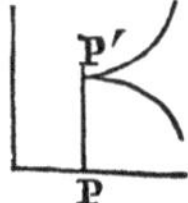

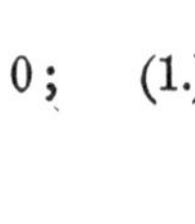

Fig. 13.

232. We come now to the geometrical construction of the general equation of the second degree, which is,

$$Ay^2 + Bxy + Cx^2 + Dy + Ex + F = 0; \quad (1.)$$

whence we derive $y = -\frac{Bx}{2A} - \frac{D}{2A}$

$$\pm \frac{1}{2A}\sqrt{\left((B^2-4AC)x^2+2(BD-2AE)x+ D^2-4AF\right)}; \qquad (2.)$$

the nature of which may be written,

$$y = A'x + B' \pm \sqrt{(C'x^2 + D'x + E')}; \qquad (2'.)$$

from (1.) we also derive $x = -\frac{By}{2C} - \frac{E}{2C}$

$$\pm \frac{1}{2C}\sqrt{\left((B^2-4AC)y^2+2(BE-2CD)y+ E^2-4CF\right)}; \qquad (3.)$$

the nature of which may be written,

$$x = A''y + B'' \pm \sqrt{(C''y^2 + D''y + E'')}. \qquad (3'.)$$

Directing our attention to equation (2.) or (2′.), we observe that its radical

$$\pm \sqrt{(C'x^2 + D'x + E')}$$

may have the value zero, at some value or two values of x, which may be found by solving the equation

$$C'x^2 + D'x + E' = 0;$$

these values found and substituted for x in

$$y = A'x + B', \qquad (2''.)$$

(which is the whole equation when the radical is zero), may give us two values of the function y and of the variable x, which must belong to the construction, in case the values are not imaginary.

Now, equation (2″.), when constructed, must be represented by a straight line. Let L L′ be this line, and the

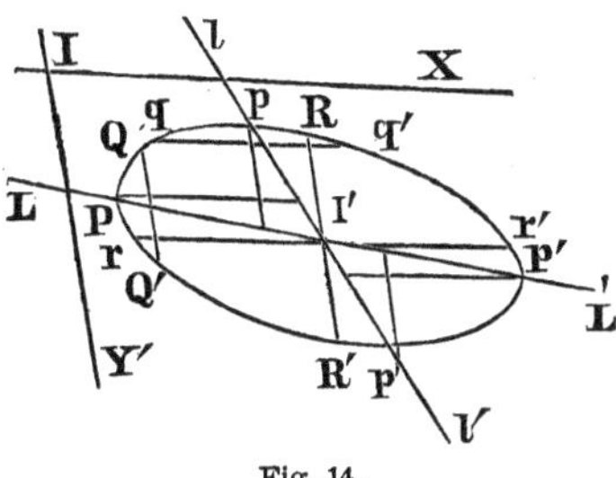

Fig. 14.

points P and P′ (Figs. 14 and 15), be determined by the values of x just found. These two values of x are,

$$x = -\frac{D'}{2C'} \pm \sqrt{\left(-\frac{E'}{C'} + \frac{D'^2}{4C'^2}\right)}.$$

If the radical term of this value of x is neither negative nor zero, we certainly have two values of x; if zero, we have one real value; if negative, we have no real value, and no geometrical construction of equation (1.) is possible.

We will suppose we have two real values of x; therefore the line we wish to construct must meet the straight line L L′ at the points P and P′.

Now, by differentiating (2′.) we have, when the radical equals zero,

$$\frac{dy}{dx} = A' \pm \frac{2C'x + D'}{2\sqrt{(C'x^2 + D'x + E')}} = \pm \infty.$$

Therefore the line sought, and necessarily meeting P and P′, must pass through these points, and for an extremely short distance must pass through them parallel with I Y, or what is the same, I Y′.

We will now proceed to regard other values of x, which

revives (2′.) in its full form, and will consider it in three respects; first, when C′ is negative; second, when C′ is positive; third when C′ is zero, that is, when in (2.) or (3.):

$$(B^2 - 4\,A\,C) < 0;$$

$$(B^2 - 4\,A\,C) > 0;$$

$$(B^2 - 4\,A\,C) = 0.$$

233. When $(B^2 - 4\,A\,C) < 0$. Regarding C′ negative in (2′.), we observe that the values of x must evidently be restricted within limits, in order that the value of the radical may not be rendered imaginary (the term $C'\,x^2$ being the ruling term.) This is so whether x be considered positively or negatively beyond the limits already found for two of its values, P and P′. Between these limits all the real values of x (and consequently y) must be contained (Fig. 14).

The straight line L L′, already located and produced if necessary, although meeting but two values of y, is nevertheless one from which all the values of the radical in (2′) are to be set off, in the two directions, parallel with I Y,′ for all possible values of y. Taking any possible value of x, we may substitute it for x in the radical, and determine other points Q and Q′, R and R′ in the line sought.

But it will be more interesting and expeditious to determine critical or singular points. Thus, the radical of

$$y = A'\,x + B' \pm \sqrt{(C'\,x^2 + D'\,x + E')},$$

calling it a separate function y', must have its own maximum and minimum, when

$$x = -\frac{D'}{2\,C'},$$

by which value of x, we determine the points R and R′,

at the greatest distance from L L′; at which points the line sought must run parallel with L L′. Again, y has its maximum and minimum separately from y'. Let us determine them at **P** and **P′**, and we know that through these points the line sought runs parallel with I X.

We have already discovered sufficient intimations of the nature of the curve line we are endeavoring to trace; that it may form a figure called an ellipse, but possibly a circle, which, however, is a particular case of an ellipse. The straight line P P′ is a diameter, and must bisect all straight lines drawn within the figure parallel to I Y′ and terminated by the circumscribing line.

Precisely the same result should we have arrived at, had we originally endeavored to construct equation (3.) as a function of y. Its form is the same, and it will be observed that the coefficient of y^2, under the radical, viz. $(B^2 - 4 A C) = C'$, is the same in both (2.) and (3.), which should be so that the ellipse should be equally indicated by each for the condition $(B^2 - 4 A C) < 0$. But instead of the line L L′, we should have $l l'$; instead of I Y, we should have I X′; instead of P and P′, we should have p and p', and any lines drawn as chords in the figure parallel to I X would be bisected by the conjugate diameter $p\ p'$, and the dif. coefs. $\frac{dy}{dx}$ and $\frac{dx}{dy}$, the reciprocals of each other in value for any given point, which of course they are by notation.

The point I′ of the intersection of these diameters may obviously be easily found at the outset from the two equations,

$$y = -\frac{Bx}{2A} - \frac{D}{2A} \qquad (4.)$$

$$x = -\frac{By}{2C} - \frac{E}{2C}, \qquad (5.)$$

where we have virtually two (functions of x) $= y$, and a

common possible value of y in each, as also of x, when the value of x and y at the intersection I′ of these diameters will be

$$x = \frac{2AE - BD}{B^2 - 4AC}, \quad (6.)$$

$$y = \frac{2CD - BE}{B^2 - 4AC}. \quad (7.)$$

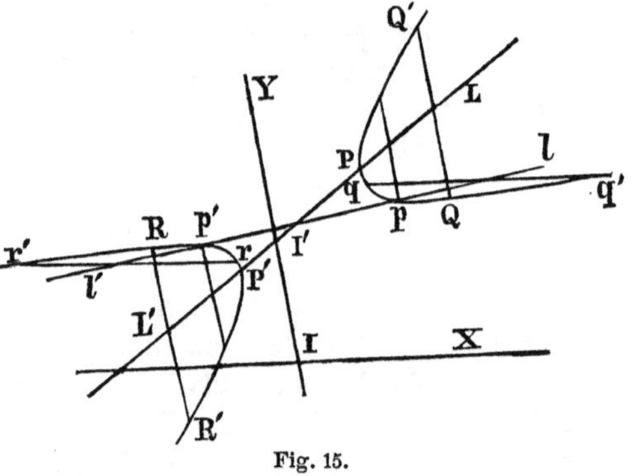

Fig. 15.

234. When $B^2 - 4AC$ is positive.

Let, as before, the line L L′ (Fig. 15) be determined, since the construction of this does not depend on the sign of $B^2 - 4AC$.

Let also certain points, P and P′, be found, at which the value of the radical becomes zero, and for each of which $\frac{dy}{dx} = \pm \infty$ as before. Therefore, through P and P′, the required line or curve will pass parallel with I Y. Let us next inquire whether it is between these two values of x now found or exterior to them, that all the real values of y are to be found, and we may at once suspect that it is exterior, for the reason that $C'x^2$, (2′.) being positive, the radical y' may have real values of unlimited greatness, positively and negatively, as well as x, while between these values of x, all values of the radical part of y, and consequently y, are imaginary. Indeed, it is quite evident that when x is sufficiently small or restricted, the terms $D'x +$ E' under the radical become ruling terms, over $C' x^2$.

Exterior to the values of x, which determine P and P′, let any value of x be assumed, and let the double values of the radical portion of y be found, for such value of x; this will determine certain points, Q and Q′, R and R′, at equal distances from L L′, as real values of the function y. In this manner we find that the line sought will be a curve consisting of two infinite, doubly symmetrical detached branches, proceeding in opposite directions. It is the hyperbola.

It is obvious that the curve in the same position might have been constructed from equation (3.), as the figure will show, by reading the previous text in small letters instead of capitals, and we thus determine p and p' etc. as we have P and P′, etc.

The lines P P′ and $p\ p'$ are called diameters of the hyperbola. If, after having assumed axes of reference, we should suppose these diameters could have any independent positions under the general equation (1.), we should mistake, for their positions being determined by equations (4.) and (5.) are such that $\frac{dy}{dx}$, from each of these two virtual functions of x cannot have opposite signs.

Now, evidently, the values of x and y at the point of intersection of these diameters for the hyperbola, are determined by equations (6.) and (7.), subject to regarding $(B^2 - 4AC)$ as negative for the ellipse, and positive for the hyperbola.

235. When $(B^2 - 4AC)$ is zero.

In this condition the term under the radical in (2.) containing x^2 becomes virtually expunged.

There can be but one value of x by which the radical in (2.) can have the value zero.

Let the line L L′, Fig. 16, be determined as in the two preceding cases. In this line let the one point P be found at which the radical equals zero. As before $\frac{dy}{dx}$ for this value is $\pm \infty$, indicating as before the course of the line sought while passing through P. Consider next whether for greater or for less values of x than that just found will the values of y be real. If greater, determine any points, Q and Q′ as before, and any other points to the right of P. This curve is a parabola, of which L L′ is a diameter.

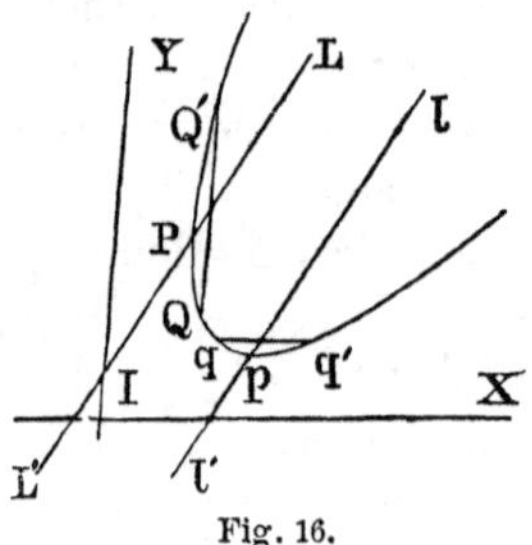

Fig. 16.

We should have constructed the same figure from equation (3.). The quantity preceding the radical would determine, as before, some other line, $l\,l'$ parallel with L L′, because the values of x and y in (6.) and (7.) become infinite. Then we should find the location of p in the same way as P, q and q' in the way as Q and Q′. If $2\,C\,D = B\,E$, the lines L L′ and $l\,l'$ become one, and the parabola becomes merged in a straight line.

In order to construct the parabola in the position required to represent the curve described by projected bodies, in agreement with the natural occurrence (Fig. 17), it will be necessary to regard the axis I Y as perpendicular to the horizon, and I X as horizontal, and to regard in equation (1.) $A = 0$, and $B = 0$, which is compatible, as it must be, with $B^2 - 4\,A\,C = 0$. Now, we have directly,

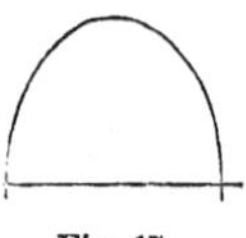
Fig. 17.

$$y = -\frac{Cx^2 + Ex + F}{D}.$$

In order to obtain the same value of y from equation (2.) we have it in this form,

$$y = -\frac{0}{0} - \infty \pm \infty.$$

which, if its meaning be sought, must be found equivalent to the previous expression. For $A = 0$, and $B = 0$, that reasoning in the foregoing text fails, which required a value or values of x to be found, at which the radical in (2.) becomes zero, since it cannot become so, unless $D = 0$, which it need not be; and if it should be, we must have,

$$y = -\frac{0}{0} - \frac{0}{0}.$$

It is a principle that no line of the second order can be intersected by a straight line in more than two points. Both branches of the hyperbola can be considered but one line, for the application of this principle. Curves of the second degree are of the simplest kind.

We have now given summarily the general construction of equations of the second degree; it may become very much simplified for particular cases. In this general construction no particular regard was needed to the particular individual signs of A or B, etc.

When the quantities, of whatever kind the units are that enter into the conditions of any problem in this treatise, can be placed in the form of the general equation of the second degree between two variables, the relation of the real values of this implicit function of either of them to the other can always be illustrated by the construction of the indicated curve, or straight line or lines.

These curves or figures are the same as those produced by the plane sections of a cone, or, in the case of the hyperbola, of two similar cones of infinite axes having their apexes in a common point, and their axes in one continuous straight line. The section of the plane through both cones always produces the hyperbola; of which two intersecting straight lines, and one straight line, are particular cases.

A section of one of the cones by the plane which if produced will not meet the other, if finite, produces the ellipse, if infinite, the parabola.

Fig. 18 is a construction for problem 27, page 85.

Fig. 19 is a construction for

$$4y^2 - 20yx + 17x^2 = 0.$$

Fig. 20 is a construction for problem 7, page 122.

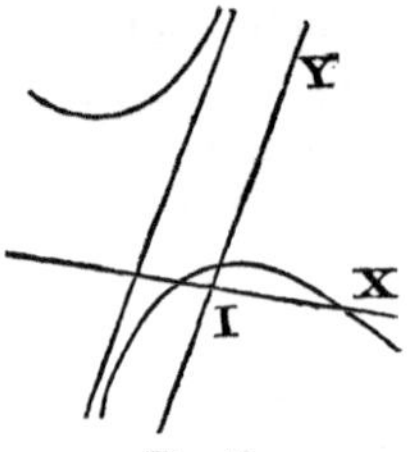

Fig. 18.

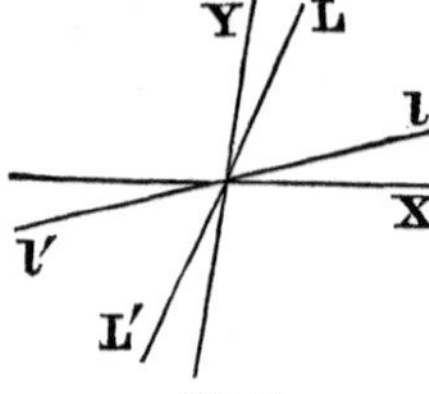

Fig. 19.

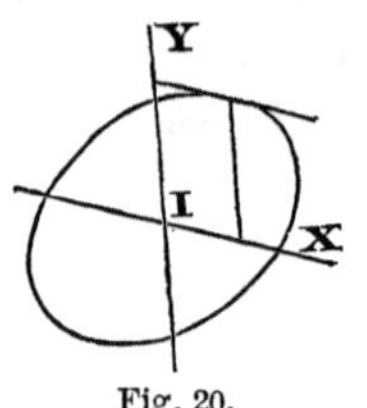

Fig. 20.

1. Required the construction of

$$y^2 - 2xy + 3x^2 + 2y - 4x - 3 = 0.$$

2. Required the construction of the equation

$$y^2 - 2xy - 3x^2 - 2y + 7x - 1 = 0.$$

3. Required the construction of the equation

$$y^2 - 4xy + 4x^2 + 2y - 7x - 1 = 0.$$

4. Required to find the roots of x and y, which in the two foregoing equations become concurrent, by constructing both equations on the same axes of reference.

236. It would be impracticable, within the limits assigned for these constructions, to undertake that of the general equation of the third or any higher degree between two variables; but we may make selections of a few functions for construction without regard to classification by degrees, and with reference for the most part to particular points of value, and to the vicinities of such points.

In the construction of the particular case of

$$y = \mathrm{A}\,x + \mathrm{B}\,x^2 + \mathrm{C}\,x^3 + \mathrm{D},$$

when the values of y are real, there always must be some position for a straight line which will intersect the curve in three points. The curve must pass through Y Y′ (Fig. 21), at the value D. There must be a point, called the point of contrary flexure, which must be at the value of x at which

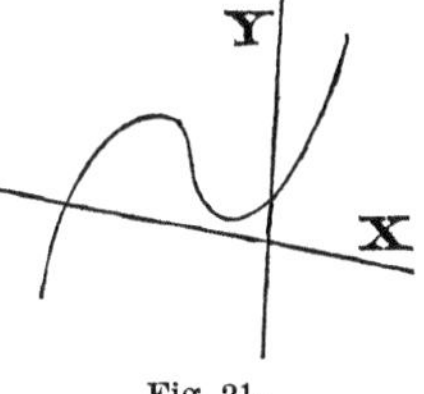

Fig. 21.

$$\frac{d^2 y}{d x^2} = 0, \text{ viz. } x = -\frac{\mathrm{B}}{2\mathrm{C}};$$

and the curve is symmetrical, with reference to this point, in opposite directions. Several problems in a previous section are based on a function of x of this nature.

In Fig. 22 is given the complete construction of

$$y^2 = a\,x^2 - x^4.$$

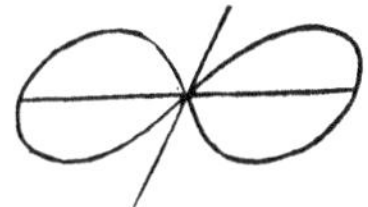

Fig. 22.

In Fig. 23 is given the construction, with the exception of infinite values, for

$$x^4 + 2\,a\,x^2\,y - a\,y^3 = 0.$$

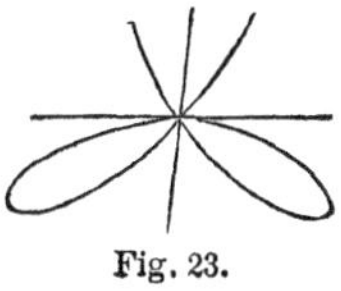

Fig. 23.

In Fig. 24 is given the construction for $x = a$, and $x = b$, and for values of x between a and b, and greater than b, for the equation,

$$y = (x - a)^2 \times \sqrt{x - b} + c.$$

In Fig. 25 is given the construction of the equation,

$$y^2 = x\,(a + x)^2 + b.$$

There is an *isolated* point of a real value of y, at $y = c$, $x = -a$; at this point $\frac{dy}{dx}$ is imaginary; which indicates that the curve has no course through that point. The value of the function for that point is drawn.

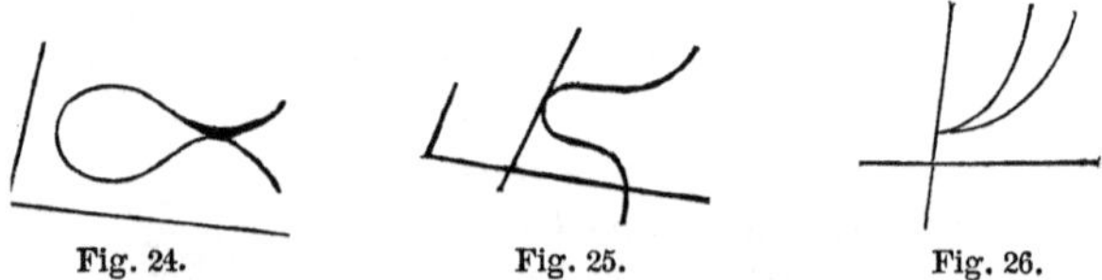

Fig. 24. Fig. 25. Fig. 26.

In Fig. 26 is given the construction for $x = 0$, $y = a$ and the vicinity, for the equation,

$$y = x^2 \pm x^{\frac{5}{2}} + a.$$

In Fig. 27 is given the construction for one point, $y = a$ and $x = b$, and the vicinity, for the equation,

$$y = a - (x - b)^{\frac{1}{3}} + (x - b)^{\frac{3}{4}}.$$

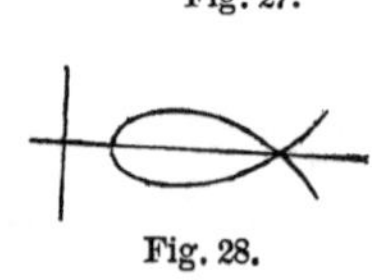

Fig. 27.

In Fig. 28 is given the construction, a being greater than b, for the points $x = a$, $x = b$, values between a and b, and greater than b, for the equation,

Fig. 28.

$$y = (x - a)\,x\,\sqrt{x - b}.$$

Let the function of x for problem 45, Section XII., be compared with this last.

237. For the geometrical illustration of the value of vanishing fractions, let the numerator $F\,x$ and the denominator $f\,x$ be constructed independently on the same axes of reference, for the value zero of each function, and for the vicinity of that value. Let N P N′ in each figure be the construction for the numerator $F\,x$, and D P D′ be that for the denominator $f\,x$, the point P being that of their concurrent value zero. Taking P O $= h$, and drawing S O, S′ O, or S S′ parallel with I Y, we shall have for that value of x next succeeding that agreeing with $F\,x = 0$, and $f\,x = 0$,

$$\frac{F(x+h)}{f\,(x+h)} = \frac{O\,S}{O\,S'}.$$

As demonstrated in the section on the subject, we have in general when $h = 0$,

$$\frac{F\,x}{f\,x} = \frac{p'}{q'} \text{ or, } \frac{p''}{q''}, \text{ or, etc.}$$

1. Fig. 29 exhibits the construction when $x = -a$, for

$$\frac{3\,(a+x)}{2\,(a+x)},$$

and its value is, by $\frac{p'}{q'}$, $= \frac{\;}{2}$; in this case it is obviously no matter whether h equal zero or not.

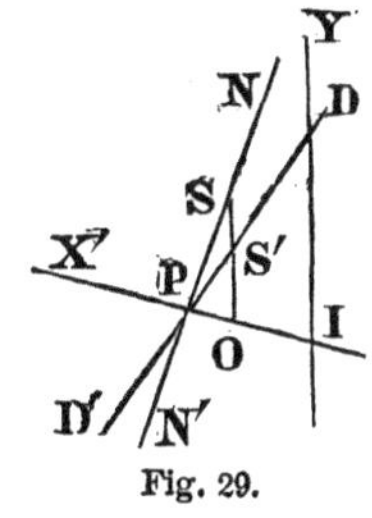

Fig. 29.

2. Fig. 30 exhibits $\frac{(a+x)^2}{a+x}$ when $x = -a$, its value being $\frac{p'}{q'} = 0$.

3. Fig. 31 exhibits $\frac{(a-x)^3}{a^3-x^3}$, when $x=a$, its value being $\frac{p'}{q'}=\infty$.

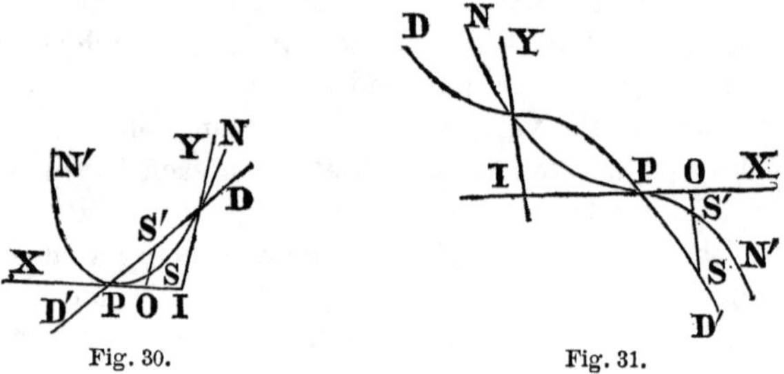

Fig. 30. Fig. 31.

4. Fig. 32 exhibits $\frac{(a-x)^2}{(a-x)^3}$, when $x=a$, its value being $\frac{p''}{q''}=\infty$.

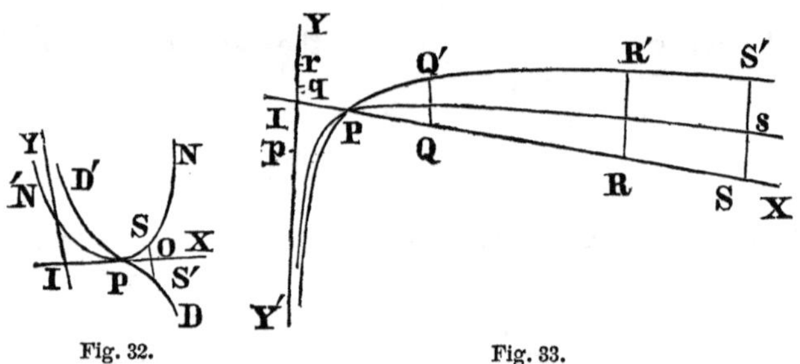

Fig. 32. Fig. 33.

In this case, since both first dif. coefs. vanish, the inspection of the figure will not show the obviousness of the value found, unless we consider that as h becomes 0, and each first dif. coef. 0, the value desired may be considered without a more particular analysis, as

$$\frac{OS'-OS}{OS-OS}=\infty.$$

238. In order to construct geometrically the associated values of logarithms and their natural numbers, let the axes of reference I X and Y Y′ intersect at I and at any angle, as in the previous constructions. Since the logarithmic function is

$$y = \log. x,$$

let the values of x, that is, the natural numbers, be taken in I X (Fig. 33), commencing at I; then may y, that is, $\log. x$, be taken in I Y when positive, and in I Y′, when negative. Take some distance I P for 1, and since log. 1 is zero for every system, the required line will meet I X at P. Now, since for the hyperbolic system

$$\frac{dy}{dx} = 1,$$

and since for values of x greater than 1, y is positive, and for values of x less than 1, y is negative, and since the dif. coef. has but one value when $x = 1$, the required line will pass through P, equally inclined to I X and Y Y′. Taking I $p = 1, =$ modulus, the straight line joining p P will be tangent to the curve, showing its course through P. For any other number or numerical quantity, I Q or I R determine the hyperbolic logarithm Q Q′, or R R′, and by the corresponding values of $\frac{dy}{dx}$ determine the course of the curve through Q′ and R′. In order to find where the tangents for any points Q′ and R′ intersect Y Y′, take in I Y, I $q =$ Q Q′ — 1, I $r =$ R R′ — 1, and q and r are the points, respectively. If I Q $= 2.71828$, Q Q′ $= 2$, and q agrees with I. If I R $= 7.38905 = (2.71828)^2$, then R R′ $= 2$, and I $r = 1$. A line passing through Q′, R′, etc., wherever they may be determined, as also through P, is the logarithmic curve for the *hyperbolic system.*

If all the values obtained for y, by the method just given, be divided by the base of the hyperbolic system, viz. 2.71828, the values of y for the *common system* will be obtained. For this system the modulus being .43429, this, instead of 1, is the amount by which any logarithm must exceed the value from I on Y Y′, at which a tangent to the curve, corresponding with such logarithm, intersects Y Y′. This curve is drawn in the figure. If I S $= 10$, then S $s = 1$, and S S′ $= 2.30258 = \frac{100000}{43429}$ = the ratio of the moduli.

The logarithmic curve may sometimes be alluded to in books when rectangular axes of reference must be understood; as in the case of the area of such curve.

239. By the construction of all functions of *one* independent variable, when this is possible, that is, when the values of them are not imaginary, and by partial construction when some values are infinite, we exhaust the availability of a plane for the representation; equally so whether the axes of reference be rectangular or not.

In order to construct a function of *two* independent variables notated $f(x, y) = z$, we may resort to space, in which solid forms are embraced; the points, lines, planes, and surfaces defined in space, however, we may project on a plane, as by Fig. 34.

Let the plane X X′ Y Y′ be intersected in Y Y′ by the plane Y Y′ Z Z′, and also in X X′ by the plane X X′ Z Z′, each intersection at any angle. If these angles be all right angles, there will be a simplification of the principle; but it should be borne in mind that they need not necessarily be so for all purposes. If x be taken of the value I N, and y of the value I O, then P is the point at which these values concur; for these values, z has a value which let P P′ parallel with Z Z′ represent. If N N′ $= h$, then P P′ becomes Q Q′. If O O′ $= k$, so that I O′ $= y + k$, then for $x =$ I N, z becomes

R R'. So that for $y + k$ and for $x + h$, we have $z = $ S S' $= F(x+h, y+k)$. Through P' and R' let a line P' R' in the plane of P P', R R' pass. Through Q' and S' let a line Q' S' in the plane of Q Q', S S' pass. Through P' and Q' let the line P' Q' in the plane of P P' and Q Q' pass. Through R' and S' let the line R' S in the plane of R R' and S S' pass. The surface expressed by P' R', P' Q', R' S', Q' S', for the vicinity of P', is determinate by the positions which z will require for its expression. The function $F(x, y) = z$, in its special character, must determine the law of this surface, for the positive and negative, single and multiplied values of z. It is obvious that the variables x and y may be independent.

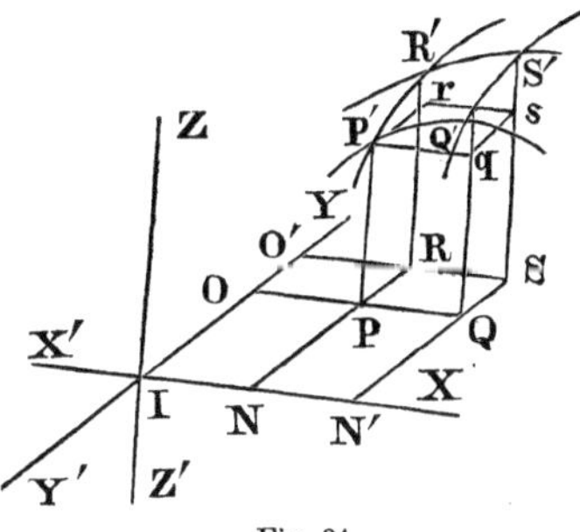

Fig. 34.

If attention is directed only to the plane P P' Q Q', the mode of representing $\frac{dz}{dx}$ is obvious; if only to the plane P P' R R', the mode of representing $\frac{dz}{dy}$ is sufficiently obvious.

The surface that is the locus of the general equation of the first degree between three variables, two of them independent, the equation being

$$A x + B y + C z = 0,$$

is a plane any how posited with reference to the three assumed planes.

The general equation of the second degree between three variables, is

$$A z^2 + B y^2 + C x^2 + D z y + E z x + F x y + G z + H y + K x + L = 0,$$

the construction of which for the locus of z, will determine a surface of such a character, that a section of it parallel with one of the planes of reference produces a curve of the second degree. Such a surface bounds a sphere; or a hyperboloid, described by a hyperbola revolving about its transverse axis;—or a paraboloid, described by the revolution of a parabola about its axis; an ellipsoid, described by the revolution of an ellipse about either its major or minor axis; or a cone, or a cylinder, the cone being a particular case of the hyperboloid; either of them any how posited.

The geometrical construction of a function of two independent variables of a higher degree than the second, will of course be by surfaces, but we can do no more than mention the methods suggested by general principles, the determination of maxima and minima for z, and of the position of the surface through any point P′ by the values of $\frac{dz}{dx}, \frac{dz}{dy}$; and refer to the analytical geometry of three dimensions, in systematic treatises. (J. R. Young's.)

A function of four or more variables can have no geometrical construction; nor one of three, when the value of z becomes imaginary.

In closing the present treatise, it is deemed worthy of remark that at least three American treatises, each partially devoted to the differential calculus, present as Taylor's theorem, both in the enunciation and in the nature of its formula, something fatally different from it, and of an unwarranted character. Let Young's Differential Calculus, one of the most scientific in the English language, be compared with these; or let the original of Brook Taylor, in the Philosophical Transactions, be consulted. There is no need of argument on this matter.

In one of the three treatises there is an apparent success in examples employing $d\,x$ as the equivalent of one minute or one second of an arc, or as being simply one in pure number. This gives a result that for the number of decimal places employed in the trigonometrical tables may be sufficiently correct; simply because one second, or even one minute happens, by the arbitrary division of the arc of a circle into certain number of degrees, minutes, or seconds, a matter utterly independent of the calculus, to be so small an arc. If the number of decimal places were increased, all these examples would be found to fail.

But having the logarithm of any number a to find log. $(a+1)$ in the manner adopted in the same treatise, by adding to log. a simply $\frac{1}{a}$, is to ignore the terms that follow in the brackets, where we give the full development for the hyperbolic system, of the direct point at issue, viz.,

$$\log.\ (a+1) = \log.\ a + \frac{1}{a}\left[-\frac{1}{2\,a^2} + \frac{1}{3\,a^3} -, \text{etc.};\right]$$

the number a must be very large that this method may not notoriously fail, for even the early decimal places of the logarithmic tables. Yet we find no intimation of a caution.

It is plain that $d\,x$ *cannot be* anything in value else than 0, since, by hypothesis and by definition, it is made so.

One of these treatises (Art. 13) undertakes to demonstrate that "if two functions are equal, their differentials are also equal;" an impossible general truth, considering these differentials as zeros, but necessarily independent. The apparent demonstration is a tautology of notation.

The same treatise, in Art. 37, undertakes to differentiate log. v by assuming, in the course of the demonstration, the development of log. $(1 + y)$, a development which must depend upon the differentiation of log. $(1 + y)$. Hence, the necessary analysis of differentiating log. v is all eliminated, and absent, and the demonstration void. Opportunities requiring criticism, as well taken as in the instances we have just presented, are copious throughout the part devoted to the differential calculus.

THE END.

ERRATA.

Page 15, line 7, *for* $\frac{Ax}{B}$ being, *read* $\frac{A}{B}$ and x.

P. 31, at foot, *dele* $-h$ *from the minuendive fraction.*

P. 45, in example 8, *for* 3^n, *read* $3n$.

P. 68, l. 16, *for the second* $-h$, *read* $+h$.

P. 73, ex. 17. *for the exponent* $\frac{3}{3}$, *read* $\frac{2}{3}$.

P. 98, l. 3, *after* range, *read* of the product of any two.

P. 120, l. 2, *for* $-3-\frac{1}{4}$ or $\frac{8}{100}$, *read* $-\frac{1}{4}$ or -3.08.

P. 140, l. 3, *for* gained, *read* lost.

P. 192, equation 2, *dele* dx *from the numerator.*

P. 197, l. 3, *for* $\frac{\mathrm{r}}{100}$, *read* $\frac{r}{100}$.

P. 201, l. 6, *draw a vinculum over* $L(\frac{9}{100})$.

P. 236, l. 6, *for* $y^2 = x(a+x)^2 + b$, *read* $y = x^{\frac{1}{2}}(a+x)+c$.

P. 244, *dele lines 5th to the 11th.*

www.ingramcontent.com/pod-product-compliance
Lightning Source LLC
LaVergne TN
LVHW010238110826
845151LV00004B/1331

9781425524623

www.ingramcontent.com/pod-product-compliance
Lightning Source LLC
LaVergne TN
LVHW010148110826
845151LV00002B/459

* 9 7 8 1 4 2 5 5 3 7 8 4 5 *